Practical Manual of Groundwater Microbiology

D. Roy Cullimore
Director
Regina Water Research Institute
University of Regina

LEWIS PUBLISHERS
Boca Raton   Ann Arbor   London   Tokyo

Library of Congress Cataloging-in-Publication Data

Cullimore, D. Roy.
  Practical manual of groundwater microbiology /
  D. Roy Cullimore.
    p.   cm.
  Includes bibliographical references and index.
  1. Water, Underground--Microbiology--Handbooks,
    manuals, etc.
  I. Title.
  QR105.5.C85 1992
  628.1'6--dc20   92-30282
  ISBN 0-87371-295-1 (acid-free paper)

LEWIS PUBLISHERS
121 South Main Street, Chelsea, Michigan 48118

PRINTED IN THE UNITED STATES OF AMERICA   1 2 3 4 5 6 7 8 9 0

Printed on acid-free paper.

Preface

"The determination of the number of bacteria in water was for a time considered of great importance, then it fell into disrepute, and the attempt was made to isolate the specific germs of diseases which were thought to be waterborne" is a quotation from William Hallock Park and Anna W. Williams in a book entitled "Pathogenic Microorganisms including bacteria and protozoa" published in 1910 by Lea & Febiger in New York. It is interesting that eight decades ago, interest in the numbers of bacteria had already waned in favor of those specific pathogens which were known to cause particular diseases of direct concern to the human species. The search for techniques to isolate and identify these organisms together with a decline in interest in the "normal" flora of waters led to a slow rate of scientific endeavor within the areas of aquatic microbiology. This mindset that the general microbial flora in waters is relatively insignificant was exacerbated in groundwaters where many believed that these waters were essentially sterile due to the filtering action of the porous media (e.g., sands, silts, gravels and porous rocks) upon the passing groundwater.

Today, it is generally recognized that microorganisms do occur naturally within the shallower groundwater systems. Their activities include the degradation of various organic pollutants which may filter into the groundwater; the plugging of flow particularly around water wells; the generation of biogases such as methane (a.k.a. natural gas) and hydrogen sulfide (a.k.a. rotten eggs smell, also known as a contributing factor to corrosive processes); changes in the chemical quality of the product waters (due to biological uptakes and releases of various chemical groups); and hygiene risk where there has been a significant contamination of the groundwater with an untreated (e.g., raw sewage) or partially treated (e.g., septic tank contents) suspension of fecal material. As in the development of all scientific knowledge, the rate of research in the area of groundwater microbiology has been biased by the demands of the users of these sources of water. The prime demands relate, from the microbiological perspective, to an assurance that there is no health risk attached to the use of the water. Essentially that the water will not contain any living entity or product from a living entity which could cause harm when the water was used in any way. For a number of decades through to today, the most accepted indicator of fecal pollution has been the coliform bacteria. Arthur T. Henrici in the book "Biology of Bacteria" published by D. C. Heath & Co. of New York in 1939 considered that these colon bacteria (coliforms) formed a satisfactory indicator group of the presence of fecal material in water. Henrici considered that "The bacteria which affect the intestines - cholera, typhoid, paratyphoid, and dysentery - are transmitted by drinking water" and usually in association with coliforms. Satisfaction with the monitoring systems and the "natural filtration activity" that groundwaters pass through while flowing in aquifers led to a lack of appreciation of the other microorganisms that may also be involved in groundwater habitats.

A number of factors have reawakened interest in the microbiology of groundwaters. Principal factors include: (1) the ongoing pollution and greater demands on surface waters causing a greater exploitation of groundwaters; (2) reducing availability of unexploited groundwater sources of adequate quality; (3) the pollution of some groundwaters with various organic and/or chemicals which renders the source unusable without some rehabilitation. Radical pollution has created hazardous waste sites where efforts are now being directed towards manipulating the microorganisms within the particular groundwater system to accelerate the processes of degradation. In surface waters such phenomena have been known as "self-purification" in which natural biological and chemical processes are active as the surface waters move downstream. Henrici in 1939 recognized self-purification of streams as being "very complex with many factors at work". Because the phenomenon was visible and conveniently recordable, self-purification of surface waters is now a well known and reasonably understood phenomenon. Self-purification of groundwaters remains however very much an enigma caught within the classic dogma "that which is out of sight is out of mind".

In writing this book, there is a major concern as there is with all newly developing sciences. The problem is the lack of attention which the science of groundwater microbiology has been given. In consequence of this, it is not possible to lean solely on the myriad of scientific and technical papers that many other branches of microbiology are fortunate to have. In its place, this book has been developed using a combination of science, observation, understanding and logic. In this case, the science is that of microbial ecology; the observation is that of watching biofouling occur both in laboratory models (microcosms) and in the field; in understanding the reasons for a particular event occurring, and the logic is the valid reasoning and correct inference relating to the behavior patterns which microorganisms may be expected to display in a groundwater habitat. Additionally, experience has been gained during the processes of developing rehabilitation techniques to recover water wells which had been severely biofouled. These experiences led to the development of a (patented) blended chemical heat treatment process which has now achieved an impressive rate of successful rehabilitations on biofouled water wells, relief wells and injection wells employed in hazardous waste site operations. Deriving from the need to determine whether or not a failing water well was suffering from biofouling, another major experience was generated. This relates to the development of an effective monitoring system to determine in a semi-quantitative and semi-qualitative manner the level of aggressivity of any microorganisms within the product (postdiluvial) water. The tests developed were the (patented) BART™ series which, in field trials, showed themselves to be potentially a more effective monitoring tool than the standard agar plate. In comparative trials, there were anomalies in the colony numbers counted on agar plates at different dilutions due to the high incidence of suspended particulates (from the biofouled zones). In this book, there is no attempt to reiterate the standard methods (Standard Methods for the Examination of Water and Wastewater, 17th edition, edited by L.S.Clesceri et al. American Public Health Association, 1989, 1624 pp. and the Supplement to the 17th edition, edited by A.E.Greenberg, et al. American Public Health Association, 1991, 150 pp.). The information for these techniques is described within the standard methods texts. Here, the objective is to allow the practitioner in the art to come to understand a little more of the mechanisms driven by microbiological activities within groundwater that can affect the predicted or actual performance of an installation which interfaces with groundwater.

The reader should recognize that the descriptions and interpretations are initial and subject to change as the science of groundwater microbiology progresses from infancy to recognition as a mature and well understood part of science. In its development, this book has been written to serve as a guide to a very new aspect of science. The science, observations, understanding and logic used are all subject to modification as more information becomes available. Many of the methodologies described within the book are new and have been used to only a limited extent in the field. As such, the information should be used for the guidance of the reader and should not be considered as a compendium of standard methods and procedures which can be used in a dogmatic manner to resolve problems relating to groundwater microbiology. For those readers who would wish to pursue in more detail the various tests and principles addressed in the book, there is a supplemental videopresentation which includes high density 3.5" floppy disks (see Appendix 3). These disks include the necessary programming (DrawPerfect v1.1, WordPerfect Corporation, Orem, Utah) to allow the presentation to run on MS-DOS compatible computers. Please note a 286 or better computer is recommended with a VGA monitor. Where it is desirable to repeat view some parts of the videopresentations, this can best be done by installing the disks as subdirectories on a hard disk where 3.2 mega of memory is available. This supplementary software can be ordered direclty from the developer together with information on future upgrading (see page x, Supplementary Software).

Information reported in this book was generated in part since 1971 when the province of Saskatchewan, through the Saskatchewan Research Council grant-in-aid program, required more information concerning the potential for iron related bacterial infestations to cause plugging in rural water wells. From this beginning, other projects were funded by such as agencies as Environment Canada, National Research Council (through direct support and the Industrial Research Assistance Program, IRAP), the Family Farm Improvement Branch of Saskatchewan, and the Natural Sciences and Engineering Research of Canada. In 1986, the American Water Resources Association organized an International Symposium on Biofouled Aquifers: Prevention and Restoration at Atlanta, GA. This symposium was funded in part by the United States Environmental Protection Agency under assistance agreement CR812759-01-0. In the same year, IPSCO of Regina partly funded along with the University of Regina, a two-day think tank on Biofilms and Biofouling. These two events brought together some of the key scientists working in the area of biofouling as it relates to ground-water. In 1990, as a partial outcome of the above meetings, a symposium on Microbiology in Civil Engineering was organized at Cranfield Institute of Technology in the U.K. This symposium was sponsored in part by the Federation of European Microbiological Societies. Paralleling this was a conference on Water Wells, Monitoring, Maintenance, Rehabilitation sponsored in part by the Overseas Development Agency of the U.K. These various events have acted as a catalyst generating a greater focus on the knowledge and ignorance that exists within the domain of groundwater microbiology. Today it is hoped that the knowledge gained will aid in the more efficient and responsible use of the groundwater resource on a global scale.

The support of the University of Regina through its ongoing funding of the Regina Water Research Institute in difficult fiscal times is gratefully acknowledged. This book would not have been possible without the very considerable and often critical advice given by Dr. Abimbola T. Abiola, George Alford, Dr. Peter Howsam, Neil Mansuy, Marina Mnushkin and Stuart Smith. Karim Naqvi is also acknowl-edged for the time he devoted to developing the software, some of which is described in the text. In particular, the dedication of Natalie Ostryzniuk in preparing this manuscript to a state where it is ready for printing with the figures and contents in their proper place, is also acknowledged.

D. Roy Cullimore, Ph.D., R.M.
Regina, Canada.

Notice

Some of the materials in this book were prepared as accounts of work sponsored by various agencies undertaken at the University of Regina (Regina Water Research Institute). Information has also been included from the body of scientific and experiential knowledge which has been generated within the general area of groundwater microbiology. This work has been reviewed by authorities familiar with the art but, it must be recognized that as a newly evolving branch of science, concepts and practices are subject to various modifications as progress is made and reported in the literature. On this basis, the Publisher assumes no responsibility nor liability for errors or any consequences arising from the use of the information contained herein.

Mention of trade names or commercial products does not constitute endorsement or recommendation for use by the Agency or the Publisher. Final determination of the suitability of any information or product for use contemplated by any user, and the manner of the use, is the sole responsibility of the user. The book is intended for informational purposes only. The reader is warned that caution must always be exercised when dealing with potentially pathogenic microorganisms, hazardous chemicals, hazardous wastes, or hazardous processes, and that expert consultation should be obtained before implementation of countermeasures for the control of potentially pathogenic microorganisms or hazardous chemical contamination.

Supplementary Software

The software referred to in the body of the text is available as a separate package from the software developer (Droycon Bioconcepts Inc., 3303 Grant Road, Regina, Saskatchewan, Canada, S4S 5H4, FAX (306) 585-3000). For more information, contact the developer directly. It includes three high density (3.5") disks. These may be viewed and used directly from the floppy drive of a 286- or better MS-DOS computer which has a VGA (or better) monitor. It is recommended that the software be installed on a hard disk for faster and more convenient use. This software will also include the drivers which will allow the various programs to be accessed directly from Windows™. Where this is installed, all of the software can be easily used through the use of a mouse clicking on the icons and drop-down menu bars which will be displayed. This package would occupy approximately 3.4 mega of memory on a hard disk.

Further information on the supplementary software can be found in the Appendices in this book. Appendix One describes the use of the BART™-SOFT program to file and interpret data obtained from the various biodetectors in that series. It will be periodically upgraded as further experience and knowledge is generated. Appendix Two examines the use of the WellRadi software to crudely project the zone of biofouling around a water well, based upon the BAQC procedure. Appendix Three describes the methods for accessing the various programs and lists the file names. In the body of the text there are references to specific programs (e.g., SHOW IRB).

Roy Cullimore was born in Oxford, England and received his PhD in Agricultural Microbiology from the University of Nottingham for studies on soil algae and developed a patented assay system using the growth of these microorganisms to assess soil fertility. Since 1968, Roy has been on the faculty at the University of Regina and is presently Director of the Regina Water Research Institute. He considers himself to be an applied microbial ecologist with a special interest in the various forms of biofouling that can be created by microorganisms. Interest in groundwater biofouling began in 1971 when asked "What do you know about iron bacteria in water wells?" and answered "Nothing, but willing to learn!" Since that time, groundwater microbiology has become a major part of the research effort. Patents emerging from this interest include co-inventor for the blended chemical heat treatment system (BCHT) for the rehabilitation of biofouled water wells; a biodetection system for specific bacterial groups in water (biological activity reaction test systems, BART™); and a new coliform test system. Working both in the laboratory and in the field, Roy has been a part of a team of researchers who have in the past decade observed many of the events which occur both during the biofouling of groundwater systems and after conventional or novel treatments have been applied. Recognition for this effort came when Roy co-chaired the American Water Resources Association International Symposium on Biofouled Aquifers held in Atlanta, Georgia in 1986. To develop some of the patents, a company (Droycon Bioconcepts Inc.) was incorporated in 1987 which now operates out of the Technology Development Facility of the University of Regina. In addition to research in the area of groundwater microbiology, Roy also is involved in the microbiology of biofouling in industrial plants, the human body and even golf courses where a lateral impermeable biofouling causes the black (plug) layer phenomenon which stresses and even kills the turfgrass. This latter project led to the development of a management system centered on the measurement of green permeability using an ascending head computer controlled infiltrometer. In moments of leisure, Roy cartoons and has had two books published and also produces artworks by etching. The only difference in these etchings is that bacteria are the artists and the environment is the "milieu."

Table of Contents

Practical Manual of Groundwater Microbiology

1

Introductory Overview

"Groundwater is an important source of water, especially in rural areas. This resource is of increasing concern to microbiologists."
Chapter 42, concept 7.[62]

Of all the environments in or upon the surface crust of this planet, the biology of groundwaters and the deep subsurface of the crust is one of the least understood. It has been a matter of convenience to consider that the biosphere does not extend down into the crust beyond shallow water wells. It was common even a decade ago to read that water from groundwater was "essentially sterile" and the potential for biological events was at the most very small and not significant. In the last decade, applied microbial ecologists have been reporting that microorganisms do indeed exist in these deeper groundwaters[2,28]. It was established that these microorganisms perform a number of very significant functions in changing the chemistry of those waters which pass through the porous and fractured media of the aquifers. These changes can affect the production from a water well, the rate at which hazardous substances such as gasoline move through the aquifer and degrade, and the quality of the product water from a well[6,50,51]. Today there is a sufficient information base to recognize these microbial events in groundwater, predict occurrences and manage the processes to the advantage of the users.

It may seem strange that while, over the last century, so many advances have been made in science and technology, so little attention has been paid to subsurface groundwater microbiology. However, the mandate of "out of sight, out of mind" applies in this case. For that matter, relatively few direct relationships have been established between the activities of the indigenous microorganisms and the structures and processes designed and managed by civil engineers. Traditionally, the main focus has been on major concerns of economic significance such as the micro-biologically induced corrosive processes which have had serious annual costs measured on a global scale in the billions of dollars for the oil industry alone.

In the last decade, the ubiquitous and dynamic nature of the intrinsic microbial activities in groundwaters is becoming recognized as inherent and essential factors involved in such events as plugging[21,22], remediation of pollution events[1,72], hydrophobicity in soils and the bioaccumulation and degradation of potentially hazardous chemicals. Some of these events can become manageable once the key controlling factors have become understood.

The microorganisms are, by and large, very simple in structure and small in size (ranging from 0.5 to 20 microns or 10^{-3} mm in cell diameter) and may commonly be spherical (coccoid) or rod-shaped (Figure 1). The impact of these organisms can create dramatic effects on the activities of what is probably the most sophisticated species, *Homo sapiens*. In recent millennia, man has learned to construct and devise many sophisticated systems which not only provide protection, safety, and transportation but now also allow rapid communication.

In today's society, engineering has created many advances which have facilitated both the intellectual and material growth of society. So rapid have been these developments, particularly in the last century, that the interaction between these systems and structures with other cohabiting species has received but casual secondary consideration. Traditionally, surface dwelling plant and animal species which are desirable strains are often "landscaped" into the environment or allowed to flourish within these manipulated structures for the direct or indirect benefit of man. Less progress has been made into the extent to which crustal dwelling organisms occupy and are active below the surface of the earth. Only today are extensive studies being initiated.

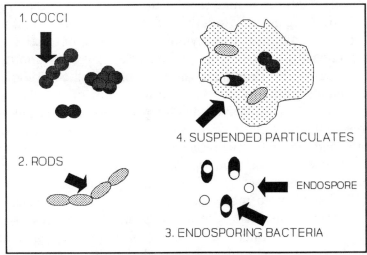

Figure 1. *Common forms of bacteria found are spherical (cocci) short rods, some of which may bear resistant endospores. Often these bacteria cohabit water retaining polymeric structures.*

CHALLENGE OF DEEP SUBSURFACE MICRO-BIOLOGY

Today, the recognition of the impact of these developments on the variety of the animal and plant species present on this planet is becoming more acutely appreciated. Such concerns raise the philosophical question that in the future protection of the earth's environment primarily as a place for the survival and proliferation of the human species, the whole nature of the crust as well as the surface of the planet is going to become subjected to engineered management structures and practices. Such a challenge falls primarily to the professional engineers.

In recognizing the need to "engineer" the crustal biological community, there needs to be understood the basic versatility that these microorganisms possess which allows them to flourish in environments within which surface dwelling creatures would rapidly die. And yet in the crust these microorganisms are able to form communities that are not only stable but also dynamic[28]. To undertake such comprehensive management practices, a very clear appreciation has to be generated on the role of the biological components as contributors to the dynamics of the crust of this planet.[12] As a surface dwelling species, it is relatively convenient for *Homo sapiens* to comprehend the role of the plants and animals since these are, by and large, either surface dwelling or flourish in the surface-waters which abound on this planet[58].

Perhaps not so recognized at this time are the roles of the microorganisms. These are insidious organisms of relatively small dimensions and simple abilities which are almost totally ubiquitous within the soils, surface- and groundwaters and even upon and within many of the living organisms which populate this planet. An example of the ubiquitous nature of microorganisms can be found in the fact that 90% of the cells in the human body are, in fact, microbial. The remaining 10% are actually the tissue cells which make up the human body as such[62]. This is possibly symbolic of the role of microorganisms throughout the surface structures of this planet. They are small, insidious, numerous and also biochemically very active.

Additionally, these microbes are able to resist, survive, adapt and flourish in some very harsh habitats. Some examples of these extreme environments[12] are:

acidophiles	growth 0.0 to 5.0 pH
alkalophiles	growth 8.5 to 11.5 pH
psychrophiles	growth range 36 to +15°C max.
thermophiles	growth range +45 to +250°C max.
aerobes	oxygen concentration from >0.02 ppm to saturated
anaerobes	no oxygen required, may even be toxic to some
barotolerant	hydrostatic pressures of 400 to 1,100 atmospheres
halophiles	growth in 2.8 to 6.2 M sodium chloride

So extreme are some of these environments that the recovery of living (viable) cells and culturing (growing) these organisms represent a set of challenges in each case. The nub of the concept of versatility within the microbial kingdom is that, provided there is liquid water and a source of energy and basic nutrients, microbial growth will eventually occur once adaptation has taken place.

One very important effect of extreme environments is that the number of species able to (1) survive and (2) flourish will decrease in proportion to the degree of extremity exhibited by the local environment. For example, the range of acidophilic species which would flourish in acidic conditions as the pH falls from 5.0 to 0.5 would diminish from as many as thirty species to as few as two.

Most of the microorganisms that have been studied originated from the surface waters, soils or the surface dependent biosphere. The widest variety of these microbes grow within a common physical domain which normally involve pressures close to 1 atmosphere; pH regimes from 6.5 to 8.5; temperatures from 8° to 45°C; redox potential (Eh) values down to -50; and salt concentrations of less than 6% sodium chloride equivalents. Outside of these domains there is a severe restriction in the range of microbial species that will flourish. A number of survival mechanisms exist ranging from either forming spores, attaching, "clustering" or forming minute suspended animation survival entities (called ultramicrobacteria).

THE GROUNDWATER ENVIRONMENT

In the groundwater within the crust, there are a number of physical and chemical factors which would begin to stress any incumbent microorganisms at deeper and deeper depths-[6,28]. This would change the nature of the dominant microbial species. However, at shallow depths where the physical factors are within the normal range for surface dwelling species, these species may continue to remain dominant. As the groundwater environment becomes deeper there are a number of factors which will influence the microbial species dominance. These factors are increases in temperature, pressure and salt concentration together with decreases in oxygen availability and the Eh potential.

In general, geothermal temperature gradients show average increase in ordinary formations of 2.13°C per 100 meters of depth[26]. Groundwater at a depth of 1,000 meters would therefore be expected to have a temperature elevation of 21.3°C and for 2,000 meters the rise would be on average 42.6°C which would take the environment up into the range where thermophilic organisms would flourish but not the surface dwelling microorganisms (mesophiles). These temperature gradient rises would be influenced by the presence of hot rock masses or unusually high or low hydraulic conductivity.

Hydraulic pressures will also be exerted to an increasing extent at greater depths, particularly below the interface between the unsaturated (vadose) zone and the saturated (water table). In general, the soil-water zone will extend 0.9 to 9.0 meters from the surface and will overburden the vadose zone. Generally in igneous rocks, the bottom of the groundwater could be at depths of 150 to 275 meters while in sedimentary rocks depths could reach 15,900 meters. In the latter case, extremely high hydraulic pressures could be exerted upon any incumbent microflora. Hydraulic pressures will be very much influenced by the hydraulic conductivity of the overburden and the degree of interconnectiveness between the various related aquifers within the system. In oceanic conditions, these factors are not significant and pressures can reach 600 atmosphere (8,820 p.s.i. or 1,280 kPa). At these higher pressures, the temperature at which water would convert to steam would also be elevated so that microorganisms could exist and flourish at superheated temperatures provided that the water was still liquid. For example, water will not boil until the temperature (°C) is reached for the pressure shown (in brackets, p.s.i.): 125°C (33.7); 150°C (69.0); 200°C (225.5); 250°C (576.6); 300°C (1,246); and 350°C (2,398). There would therefore appear to be a strong probability that there may be a series of stratified microbial communities within the crust which are separable by their ability to function under heavier pressures in more concentrated salts at higher temperatures under extremely low (-Eh) redox potentials (e.g., -450 mV).

For these microorganisms to survive and flourish in extremely deep subsurface environments, the geologic formations must provide openings in which the water (and the organisms) can exist. Here the microbes will, in all probability, grow attached to the surfaces presented within the openings. Typical openings include intergrain pores (in unconsolidated sandstone, gravel, and shale), systematic joints (in metamorphic and igneous rocks, limestone), cooling fractures (in basalt), solution cavities (in limestone), gas-bubble holes and lava tubes (in basalt) and openings in fault zones. All of these openings provide surfaces large enough to support such attachment if the cells can reach the site. Such a restriction would be relatable to the size of the microbial cells in relation to the size of the openings.

Porosity is usually measured as the percentage of the bulk volume of the porous medium that is occupied by interstices and can be occupied with water when the medium is saturated. For coarse to fine gravel, the percentage porosity can range from 28 to 34% respectively. Porosities for sand tend to range from 39 to 43%. Fine and medium grain sandstone possess porosities from 33 to 37% respectively. A tightly cemented sandstone would have a porosity of 5%. However rocks tend to be porous structures with pore sizes large enough to accommodate bacterial cells which, in the vegetative state, have a cell diameter of between 0.5 and 5 microns. When these cells enter a severe stress state due to starvation or environmental anomalies, many bacteria will shrink in size to 0.1 to 0.5 microns and become non-attachable. Such stressed survival cells are referred to as ultramicrobacteria and are able to passage through porous structures for considerable distances. Recovery is dependent upon a favorable environment being reached which would allow growth and reproduction. These ultramicrobacteria therefore have the potential to travel through porous media and remain in a state of suspended animation for very prolonged periods of time. Even normal vegetative cells (e.g., those of *Serratia marcescens*) have been shown to pass through cores of Berea sandstone (36.1 cms) and limestone (7.6 cms) during laboratory studies. There is indeed a growing body of knowledge supporting the ability of bacteria[55] to invade and colonize porous rock formations. Viruses have also been recorded as being able to travel through porous structures[35].

Fractures in rocks tend to have much larger openings than pore structures and can therefore form into potential groundwater flow paths which could act as focal sites for microbial activities (Figure 1). This heightened activity would result in part from the passage of groundwater containing potentially valuable dissolved and suspended chemicals (e.g., organics, nutrients, oxygen). In addition, the flow paths would allow the transportation of the microbial cells and also the dissolution of waste products. Microbial mobility through fracture flow paths in rocks can be very fast. In 1973, bacteria were recorded to travel at rates of up to 28.7 m/day through a fractured crystalline bedrock[55].

Little attention has been given to the scale of microbiological activity present in soils or the crust of the earth. It is only in the last decade that attention is being paid to the potential size of the biomass occurring in both soils and crustal elements. Present extrapolations (Figure 2) indicate that the biomass is indeed very significant in relation to the biomass found on the surface of the planet. Further researches will undoubtedly allow a more precise comparison of these various parts of the biosphere.

MICROBIAL MECHANISMS OF COLONIZATION AND GROWTH

There arises the need to understand the mechanisms by which these microbial invasions of porous and fractured structures can occur in both saturated, capillary unsaturated and vadose zones. It has perhaps traditionally been thought that, since the groundwaters are essentially "sterile", any microbial problem in a water well must have been the result

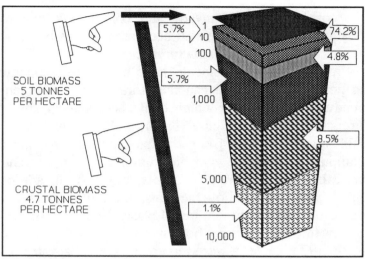

Figure 2. *Recent findings[12] and research[28] indicate that a microbially dominated biosphere extends into the crust. Soil biomass is based on the A horizon, crustal biomass to 10,000 meters.*

of a contamination of the well during installation with microorganisms from the surface environment. In reality, while such contaminations do occur, there is also the possibility that microbes may migrate towards the well as an "inviting" new environment to colonize. The advantages are the potential presence of oxygen (diffusing down both inside and around the well to support the aerobic microorganisms) and a more rapid and turbulent flow of groundwater as it approaches and enters the well which will deliver a greater potential quantity of nutrients to the well in comparison to the bulk aquifer. It is interesting to note that the maximum microbial activity tends to occur at an interface where the oxygen is diffusing one way (i.e., away from the well) and nutrients are moving towards the well (i.e., from the aquifer formations). In general, such focused microbial activities occur under conditions where the environment is shifting from a reductive regime to an oxidative one as a result of the movement of water towards the well. Maximum microbial activity commonly occurs where the environmental conditions are just oxidative (Eh potential, -50 to +150). A cylinder of microbial growth can therefore be envisaged as the biozone of maximal microbial activity at this reduction-oxidation (REDOX) fringe.

One major step in the development of an understanding is to appreciate that the growth of different species of microorganisms often occurs within common consortial structures where the various species co-exist together. These structures are bound with water retaining polymeric matrices to form biofilms often attached to inert surface structures. These biofilms, more commonly known as slimes or plugs, act as catalysts which will then generate such events as corrosion (through the generation of hydrogen sulfide and/or organic acids), and bioaccumulation of selected inorganic and organic chemicals travelling generally in an aqueous matrix over the biofilm[16]. Biodegradation may also occur. Once the biofilm has enlarged sufficiently, a total plugging of the infested porous medium can sometimes occur[25]. These latter events hold a tremendous potential for the control of the many hazardous wastes which have inadvertently entered the surface and ground waters, soils and rock strata on this planet. By coming to understand the methods by which these microbial activities can be utilized, it will become possible to include these manipulations of microbial processes as an integral part of engineering practices.

Microbial events[6,22,45,50,51,72] can include corrosion initiation, bioimpedance of hydraulic or gaseous flows (e.g., plugging, clogging), bioaccumulation of chemicals (e.g., localized concentration of heavy metals, hydrocarbons and/or radionuclides within a groundwater system), biodegradation (e.g., catabolism of potentially harmful organic compounds), biogenesis of gases (including methane, hydrogen, carbon dioxide, nitrogen which can lead to such events as the fracturing of clays, displacement of water tables and differential movement of soil particles), and water retention within biofilms (which, in turn, can influence the rates of desiccation and/or freezing of soils). These events occur naturally within the biosphere as microbiologically driven functions within and upon the crust of the planet (Figure 3).

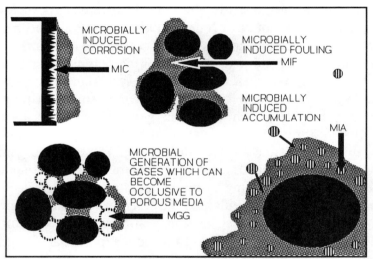

Figure 3. *Common forms of microbial events (MIC, MIF, MGG and MIA) can occur within groundwaters and affect the integrity of engineered systems.*

1. **MIF** (microbially induced fouling) where the microorganisms begin to generate a confluent fouling of the surfaces of the porous media, particularly at the reduction-oxidation interface. As these biofilms expand and interconnect, so the transmissivity of water through the system becomes reduced (plugging).

2. **MIA** (microbially induced accumulation) will occur at the same time and will involve a bioaccumulation of various ions such as metallic elements in the form of dissolved or insoluble salts and organic complexes.

3. **MGG** (microbial generation of gases) with the maturation of the biofilm and extension of anaerobic growth, gas formation is much more likely. Gases may be composed of carbon dioxide, methane, hydrogen and nitrogen and may be retained within the biofilm as gas filled vacuoles or in the dissolved state. Radical gas generated biofilm volume expansion could radically reduce hydraulic transmissivity through the biofouled porous structures.

4. **MIC** (microbially induced corrosion) poses a major threat to the integrity of structures and systems through the induction of corrosive processes. Where the biofilm have stratified and/or incorporate deeper permanently anaerobic strata there develops a greater risk of corrosion. Such corrosive processes may involve the generation of hydrogen sulfide (electrolytic corrosion) and/or organic acids (solubilization of metals).

5. **MIR** (microbially induced relocation) occurs where the biofilm growth and biological activities associated with these microbial activities cause a shift in the redox potential or the environmental conditions in such a way that there is a relocation of the biofilm. Here, the biofilm begins to sheer and slough away carrying some of the biomass and accumulates in the flow. Relocation may also be a managed function when the environmental conditions are artificially manipulated.

At this time, the focus of attention has been directed to the use of specific strains of microorganisms which have been selected or manipulated to undertake very specific biochemical functions under controlled conditions. While the biotechnological industry has achieved some successes in producing useful products, it has failed to fully recognize the potential for manipulating the natural microflora to achieve a "real world" conclusion (such as the bioremediation of a nuisance or potentially toxic chemical). These events occur naturally around water wells and may influence the resultant water quality more than is generally recognized.

MANAGEMENT OF MICROBIAL ACTIVITIES

Microorganisms are influenced by a range of physical and chemical constraints which are unique to each strain[12]. Generalizations can therefore be made only for major groups of bacteria, but natural variations will always occur. Before these variables can be considered in depth, the form of microbial growth has to be considered. A popular conception of microorganisms is that: (1) these are cells usually dispersed in water and (2) many may be able to move (i.e., be motile). In actuality, microbial cells can commonly be found in three states. These are: **planktonic** (dispersed in the aqueous phase), **sessile** (attached within a biofilm to a solid, usually immobile surface), and **sessile particulate** (incumbent within a common suspended particle shared with other cells, Figure 3).

In soils and waters, the vast majority of the microbial cells are present in the sessile phases rather than in the planktonic. Sessile microorganisms are normally found within a biofilm created by the cells excreting extra-cellular polymeric substances (ECPS)[16]. These polymers act in a number of ways to protect the incumbent cells. This action includes the retention of bonded water, accumulation of both nutrients and potential toxicants, providing a structural integrity (which may include gas vacuoles). Upon the shearing from the biofilm, the particulate structures afford protection to the incumbent cells. These sheared suspended particulates migrate until reattachment and colonization of a pristine (econiche) surface.

Recognizing that to control these microbial activities, there will be a need to exercise control over the rate of selected microbial activity within the defined environment, it will become essential to build a management structure that will allow the control of all of the major factors of influence.

These can primarily be subdivided into four major groups which are: (1) physical, (2) chemical, (3) biological, and (4) structural. Each group not only very significantly reacts with the other major groups but also includes a series of significant subgroups each of which can also become dominant on occasions.

For the individual microbial strain or consortium of strains, there is a set of environmental conditions within the environment that relates to these four major factors. These allow particular definable econiches to become established. Such microbial associations are often transient in both the qualitative and quantitative aspects due largely to the dynamic and competitive nature of the incumbents. In chronological terms, there is a very limited understanding of the maturation rate of these natural consortia. For example, what is the length of time that an active microbial biomass can remain integrated within an econiche positioned within the totally occluded zone around a plugged water well? This has not been addressed. Nor has the question of how long would the period be before a water well could become biologically plugged.

Many consortia of fungi (molds) and algae can form very specific and unusual forms of growth[12,62]. For example, surface-dwelling lichens grow at rates as slow as 0.1 mm per year often in an environment (i.e., the surface of a rock) in which nutrients are scarce and water rarely available. Here, the microbial activity is at a level where maturation may be measured in decades rather than days, weeks or even months. This illustrates a condition in which microorganisms have adapted to an apparently non-colonizable hostile environment and yet formed an econiche within which growth of a consortial nature can flourish for perhaps centuries.

Of the physical factors known to impact on microbial processes, the most recognized factors are temperature, pH, redox and water potential[22]. Functionally, microbial processes appear to minimally require the presence of temperature and pressure regimes which would allow water to occur in the liquid form. Low temperatures cause freezing in which ice crystals may form a matrix around any particulates containing microorganisms and the liquid (bound) water-retaining polymerics. Water may then be expressed from the liquid immobilized particulates to the solid ice fraction leading to a shrinking of the particulate volume until a balance is achieved between the bound liquid water and the ice. Such low temperatures inhibit biochemical activities causing a state of suspended animation to ensue. At higher temperatures, water can remain in a liquid phase under increasing pressures up to at least 374°C, rendering it a theoretical possibility that microbial activities could occur in such extreme environments.

Frequently, microorganisms in suspended "sessile" particulate structures can adjust the density of these entities causing them to rise or fall within a water body (Figure 4). Elevation (a lowering of density) may be caused by the formation of entrapped gas bubbles or the releasing of some of the denser bioaccumulates. Density may increase (particle sinks) where either water is expressed from the particle causing the volume to shrink or the bioaccumulation of dense material occurs.

Normal surface dwelling microorganisms[3], however, generally function over a relatively limited temperature range of 10° to 40°C with optimal activities occurring at a point commonly skewed from the mid-point of the operational range of growth. These optima are to some extent different for each strain of bacteria and therefore any temperature changes can cause shifts in the consortial structure of microorganisms.

Optimal and functional ranges of activity also exist when the microorganisms interact with the environmental pH

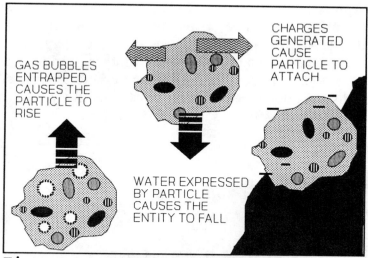

Figure 4. *Particles can have the buoyancy microbially manipulated upward (left, gas generation), downward (center, water expression), and attaching (right, electrically charged).*

regime. Most microbes function most effectively at neutral or slightly alkaline pH values with optima commonly occurring at between 7.4 and 8.6. Acidoduric organisms are able to tolerate, but not necessarily grow at much lower pH values such as 1.5 to 2.0 pH units but are generally found in specialized habitats. Where biofilms have been generated within an environment, any pH shifts may be buffered by the polymeric matrices of the biofilms to allow the incumbent microorganisms to survive and function within an acceptable pH range being maintained by the consortial activities while surrounded by a hostile pH.

Redox has a major controlling impact on microorganisms through the oxidation-reduction state of the environment. Generally, an oxidative state (+Eh values) will support aerobic microbial activities while a reductive state (-Eh) will encourage anaerobic functions which, in general, may produce a slower rate of biomass generation, a downward shift in pH where organic acids are produced, and a greater production potential for gas generation (e.g., methane, hydrogen sulphide, nitrogen, hydrogen, carbon dioxide). Aerobic microbial activities (Figure 5) forming sessile growths are often noted to occur most extensively over the transitional redox fringe from -50 to +150 mV.

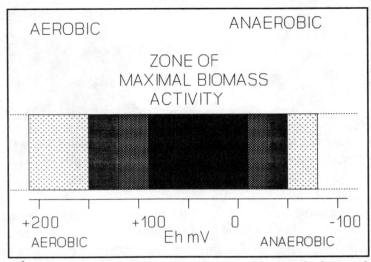

Figure 5. *In groundwater systems, the microbial growth (biomass) will often focus at the reduction - oxidation interface at Eh values from +150 to -50 mV.*

Chemical factors influencing microbial growth can be simply differentiated into two major groups based upon whether the chemical can be stimulative or inhibitory to the activities within the biomass. The stimulative chemicals would in general perform functional nutritional roles over an optimal (most supportive) concentration range when the ratio of the inherent nutritional elements (e.g., carbon, nitrogen, phosphorus) are in an acceptable range to facilitate biosynthetic functions. These concentrations and ratios vary considerably among microbial strains and form one of the key selective factors causing shifts in activity levels within a biomass. In the management of a biological activity, it is essential to comprehend the range and interrelationships of any applied nutrients in order to maximize the biological event that is desired.

Inhibitory chemicals can generate negative impacts on the activities of the biomass, taking the form of general or selective toxicities. For example, heavy metals tend to have a generalized type of toxicity while an antibiotic generated by one member of a consortium may be selectively toxic to some transient or undesirable microbial vectors. It is surprising that the biofilm itself may tend to act as an accumulator for many of these potentially toxic elements which may become concentrated within the polymeric matrices of the glycocalyx (structure forming the biofilm). When concentrated at these sites, these toxicants can form a barrier to predation. At the same time, these accumulates are presumably divorced from direct contact with the incumbent microorganisms.

In recent scientific progress, some new ideas are being evaluated using biological vectors which can be used to directly control the activities of a targeted group of organisms. Traditionally, the concept of control focused on the deliberate inoculation of a predator (to feed upon) or pathogen (to infect) the targeted organisms with some form of disease. In the last decade, there have been impressive advances in the manipulation of the genetic materials, particularly in microorganisms, which has generated new management concepts. These involve the inoculation of genetically modified organisms which have been demonstrated, in the laboratory setting, to have superior abilities (e.g., to degrade a specific nuisance chemical) compared to microbial isolates taken from the environment. The objective of these biotechnologies is that superior functions can be transplanted into the natural environment and achieve equally superior results to that observed in the laboratory. Such concepts still have to address the ability of the inoculated genetically modified organism to effectively compete within the individual econiches (some of which may be as small as 50 microns in diameter) with the incumbent microflora and still perform the desired role effectively.

Another major factor influencing any attempt at microbial manipulation within a natural environment are the impedances created by the physical structures between the manipulator and the targeted econiche. These structures also influence the environmental characteristics of that habitat. Manipulations may be relatively easy to achieve where the target activity is either at a visible and accessible solid or liquid surface/air interface or within a surface water (preferably non-flowing). Subsurface interfaces involve a range of impedances not only to the manipulator but also to the incumbent organisms. In order to manipulate a subsurface environment, there are two potential strategies. The first is physical intercedence where a pathway is created to the targeted site by drilling, excavation, fracturing or solution-removal processes.   Here, there will be some level of contamination of the target zone with materials and organisms displaced by the intercedence processes.   As an alternate method, the second strategy involves a diffusive intercedence where the manipulation is performed remotely by diffusion or conductance to achieve the treatment conditions towards the targeted zone.   Such techniques may involve the application of gases (e.g., air, methane), solutions (containing optimal configurations of the desired chemicals to create appropriate conditions through modifying the environment), suspensions of microorganisms in either vegetative, sporulated or suspended animation states (such as the ultramicrobacteria), thermal gradients (through the application of heat or refrigeration) or permeability barriers that will cause a redirection or impedance of hydraulic conductivity flows in such a way as to induce the desired effect.

Diffusive intercedence can offer simpler and less expensive protocols but may be more vulnerable to the effects that any indigenous microflora may have upon the process. These effects could range from the direct assimilation of the nutrients, modification to the physical movement of the materials being entranced, to the releases of metabolic products which may support the desired effect.

Engineering structures and processes in an environment free from microorganisms may be relatively easy to design and manage. On this planet, the microbial kingdom forms a diverse group of species which occupies a wide variety of environments. When engineering projects are undertaken, these indigenous organisms respond in ways that are often subtle and often these go undetected. With an improvement in the ability to recognize and control microbial processes, there should develop not only ways to manage these microbial events but also to recognize the benefits as well as the risks of having such events occur.

2

Concepts

Marcus Aurelius Antoninus wrote in his meditations published in the second century A.D. (Volume V,23), "For substance is like a river in a continual flow, and the activities of things are in constant change, and ... causes work in infinite varieties; and there is hardly anything which stands still" and (Volume IX,36) "The rottenness of matter which is the foundation of everything! Water, dust, bones, filth, or again, marble rocks, the callosities of the earth; and gold and silver, the sediments; and garments, only bits of hair; and purple dye, blood; and everything else is of the same kind. And that which is of the nature of breath, is also another thing of the same kind, changing from this to that" referenced many of the microbiological events which, at various rates, recycle the elements found within and upon the surface crust of this planet.

Much of the microbiological activities associated with these events occur within complex consortial communities formed within surface-dwelling biofilms. Popularly referred to as "slimes", these communities often stratify, act as chemical accumulators of both nutritional and toxic chemicals, and cause plugging and corrosion. These events have been witnessed over the course of history. Darwin, for example, writing in 1845 on "The Voyage of the Beagle" references the gold mine at Jajuel, Chile and reports that the tailings piles left after the gold has been separated were being heaped, after which "a great deal of chemical action then commences, salts of various kinds effloresce on the surface, and the mass becomes hard (after leaving for two years and washing) ..it yields gold.... repeated even six or seven times...there can be no doubt that the chemical action...liberates fresh gold". Here, Darwin was reporting a mine ore leaching process which was, for many years, thought to be a chemical process but is now recognized to be biological and to involve the formation of biofilms. Ionic forms of gold are now known to be microbially amended to the elemental state [62] which may become deposited in the cell walls.

Within such growths, there are incumbent strains of microorganisms which compete with each other for dominance in the maturing biofilm and cause secondary effects, for example, plugging, corrosion and bioaccumulation.

The terms biofilms and biofouling can be defined as relating to the cause and effect, respectively, of any microbial physical and/or chemical intercedence into a natural or engineered process by a biologically generated entity. The structure of the biofilm is defined in terms of the incumbent qualitative and quantitative aspects with particular emphasis on the phases of maturation through to a completion of the life cycle (e.g., occlusion of a porous medium, total plugging).

Major environmental factors influencing the rate of growth of a biofilm are the C:N:P ratio, temperature, redox potential, and pH. These are generally considered to be the particularly important factors. Dispersion of biofilms would allow the consortial organisms to migrate to fresh, as yet uncolonized, econiches as either suspended particulates (due to shearing) or non-attachable ultramicrobacteria (generated as a response by many microorganisms to severe environmental stresses). The net effects of biofouling are usually considered in terms of causing various plugging, corrosion, bioaccumulation, biodegradation, gas generation activities, and also causing changes in the characteristics of water retention and movement in porous media such as those found in soils, surface waters and groundwaters, and in engineered structures. Often biofilms are more commonly associated with corrosion and plugging events than with the hygiene risk factors.

Traditionally, the major concern in relationship to groundwater microbiology was to recognize any potential health hazard which may be present[11,26,31,64,80,84]. These hazards were generally viewed as significant where there was a distinct presence of coliform bacteria identifiable at the significance level of at least one viable entity per 100 mL. Coliform presences are linked to the potential for direct (fecal coliform) or indirect (total coliform) contamination of the water by fecal materials. In practice, the most common concern relates to contamination with domestic wastewaters, manure leachage, liquid waste-holding tanks and industrial discharges.

Today there is a growing realization that there is a wider range of microorganisms which can present a significant health risk and yet may not be associable with the presence of indicator organisms for fecal material(e.g., coliforms) [6,7,22,50,54]. In many cases, these organisms are nosocomial pathogens in that they are normal inhabitants of particular environments and yet are also able to cause clinical health problems in stressed hosts. The form of stress can vary from nutritional (e.g., starved), physiological (e.g., exhaustion or surgical aftereffects) or immunological (e.g., suppressed). Nosocomial pathogens are commonly present within natural environments and are often difficult to distinguish from the natural background microflora. This is particularly the case where the nosocomial pathogen forms

a part of the consortium within a biofilm growing attached
to the various surfaces in the econiche.

MICROBIAL DYNAMICS OF BIOFILMS

Biofilms are dynamic[16] and there is an ongoing competition between the incumbent microbial strains which may
result in three effects. Firstly, the range of strains which
may be present can become reduced or stratified. Qualitative reductions will occur where one or more of the strains
become more efficient than the competing organisms and
gain dominance. Young natural biofilms can often contain
as many as 25 or 30 strains which will, during the process
of maturation, become reduced in number to as few as five
strains or less[50]. Another response commonly seen in a
maturing biofilm is the stratification of the biofilm in which
the primary layering differentiates an exposed aerobic
(oxygen present) stratum overlaying and protecting lower
anaerobic (oxygen absent) layers. These lower surface
interfacing layers often will biologically generate hydrogen
sulfide and/or organic acids which may then initiate the
processes of corrosion.

A second major effect relates to the incumbency density
of cells within the biofilm which may fluctuate with environmental conditions. This incumbency density (*id*, as viable
units per volume of particulate material) may be measured
by undertaking an enumeration of the viable units by a technique such as extinction dilution spreadplate estimation of the
colony forming units per mL (cfu/mL). Concurrently, the
volume of the biofilm can be estimated using the dispersed
particulates from the biofilm and measuring the volume with
a laser driven particle counter.   The volume can be
summarized as mg/L of total suspended solids (TSS). From
the estimation of the number of viable units per mL (vu) and

the volume of the suspended solid particulate material per mL (vp, computed as the TSS divided by 1,000), it becomes possible to project the incumbency density (id):

$$id \quad = \quad vu \ / \ vp$$

Here, the id would commonly be recorded as the incumbency density of colony forming units per microgram (10^{-6} g) assuming that the particulate density was identical to that of water (i.e., 1.0). An example of the effect of stress on the id would be where a gasoline spill had occurred in a groundwater system. Stressed biofilms, for example, impacted upon by a significant bioaccumulation of gasoline (e.g., 16 mg/mL biofilm) have been recorded as having id as low as 20 cfu/10^{-9} g [Cullimore, D.R. and A. Abiola, 1990, Biofouling of Groundwater Systems in (ed. D.H. Waller) Emerging Research Problems and Methodologies Related to On-site Sub-surface Wastewater Disposal, Pub. Tech. Press. C.W.R.S. Halifax, Nova Scotia, 125-138]. When nutrients were concurrently applied, the id values increased by as much as one thousand fold (i.e., 20,000 cfu/10^{-9} g).

Competition between the incumbent microbial strains in a maturing biofilm in response to environmental conditions can lead to a third effect which is related to the degree of imposed stress. Responses can occur at the individual cell level or to the biofilm itself. Where stress impacts upon an individual cell as a result of trauma, it can cause a response. These traumas can be created by a shift in one or more environmental factors which no longer allow growth. Such events can include a shortage of nutrients or oxygen, a build up in toxic end-products of metabolites or antibiotics, and losses in structural integrity in the biofilm. In general, cells under extreme stress will void nonessential water and organic structures, shrink in volume with a tenfold reduction in cell diameter, generate an electrically neutral cell wall (to reduce attachability), undergo binary fission and enter a phase of suspended animation. These minute non-attachable cells (Figure 6) can remain viable for very long periods of time as ultramicrobacteria. In this state, bacteria can move with any hydraulic flows as suspended animation particulates until the environmental conditions again become favorable. Once such facilitating conditions recur, the cells again expand to a typical vegetative state and colonize any suitable econiches. A newly developed water well can provide a suitable econiche for the transient aerobic ultramicrobacteria passing through the groundwater.

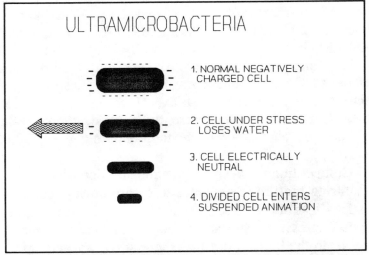

ULTRAMICROBACTERIA

1. NORMAL NEGATIVELY CHARGED CELL

2. CELL UNDER STRESS LOSES WATER

3. CELL ELECTRICALLY NEUTRAL

4. DIVIDED CELL ENTERS SUSPENDED ANIMATION

Figure 6. *When under stress, bacterial cells can undergo a series of events (descending column) leading to the generation of two very small ultramicrobacterial cells.*

BIOFILM MATURATION

General concepts of bacterial growth may often envisage the cells as simply floating or "swimming" in water. In reality, bacteria tend to attach (Figure 7) onto surfaces and then colonize the surface entirely to form a biofilm. The bacterial cells tend to be negatively charged and are attracted to positively charged surfaces. When this happens, the bacterial cells become anchored to the surface by extending polymers (long chained molecules) which make the primary attachment. Subsequently the cells will reproduce and colonize the surface using such mechanisms as "jumping", "tumbling" or they may simply clump to form a micro-colony on the surface.

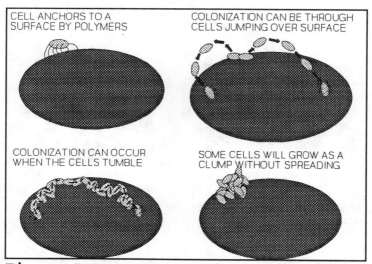

Figure 7. *Bacteria are attracted to and attach to surfaces. Subsequent reproduction may involve colonization by either "jumping", "tumbling" or the cells may simply clump together.*

Biofilms may respond to stress as a normal part of the maturation cycle[50,51]. From studies undertaken using bio-fouled laboratory model water wells (1 L capacity), a sequence of events has been recorded following a basic pattern for the development of the iron related bacterial biofilms.

Once the colonization of a surface has begun, one of the common early events is the rapid spread of microorganisms over all of the surfaces that present a "friendly" environment. These organisms grow within a water-retaining biofilm (Figure 8). Initially, this biofilm will contain a

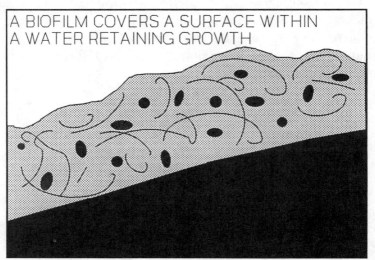

Figure 8. *Bacteria grow in mixed consortia containing many different species within a biofilm attached to a solid surface (black). Water-retaining polymers form the structure of the "slime".*

randomized mixture of a wide variety of microorganisms. Over time, however, these organisms will either stratify into distinct parts of the biofilm, become dominant or be eliminated from the consortium forming the biofilm. A maturing biofilm is sometimes referred to as a glycocalyx.

Biofilms form a biological interface with the water passing over the surface. As the water passes over, chemicals are extracted and concentrated within the biofilm (Figure 9). These chemicals may be grouped into two major categories: nutrients and bioaccumulates. Nutrients are

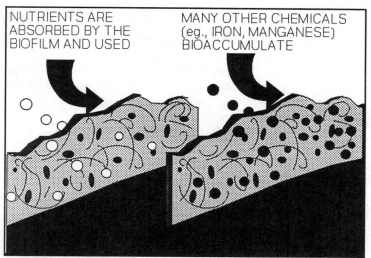

Figure 9. *Biofilms (shown sectioned) will absorb and utilize nutrients (left, open circles) but may simply bioaccumulate other chemicals (right, black circles) such as iron during the lifetime of the biofilm.*

utilized by the microorganisms for growth and reproduction and so the concentrations usually do not continue to build up indefinitely. Bioaccumulates, however, are not used, by and large, and so accumulate with the polymeric structures of the biofilm. Commonly accumulates include non-degradable organics and various metallic ions such as those of iron, manganese, aluminium, copper and zinc. The role of these bioaccumulates would appear to be relatively "passive" but it is generally believed that these compounds reduce the risk of predation by scavenging organisms.

In this maturation cycle, a number of distinct events could be observed. These can be categorized into a number of sequential phases after the formation of a confluent biofilm.

(1)   rapid biofilm volume expansion into the interstitial spaces with parallel losses in flow;

(2)   biofilm resistance to flow next declines rapidly causing facilitated flows which can exceed those recorded under pristine (unfouled) conditions;

(3)   biofilm volume compresses, facilitated flow continues;

(4)   biofilm expands with periodic sloughing; causes increases in resistance to hydraulic flow (intercedent flow). There is a repeated cycling with a primary minor biofilm volume expansion followed by secondary increased resistance to flow ending in a tertiary stable period (see SHOW BIOFILM-2);

(5)   interconnection between individual biofilms now generates a semi-permeable biological barrier within which free interstitial water and gases may become integrated to form an impermeable barrier (plug).

The speed with which these phases are generated is a reflection of the environmental conditions occurring in the laboratory or field setting (Figure 10). For example, in the laboratory, phases 1 to 3 can be accomplished within ten days when generating a black plug layer using sulfur coated urea fertilizer in high sand content columns. Pulses occur in phase 4 in which the biofilm expands slightly, becomes unstable increasing resistance to flow and then stabilizes. In a model biofouled water well these pulses can take 25 to 30 days per cycle. In producing wells, cycles as short as ten to twelve days have been recorded.

During the phase 4 increases in resistance to flow, it can be projected that the polymeric matrices forming the biofilm

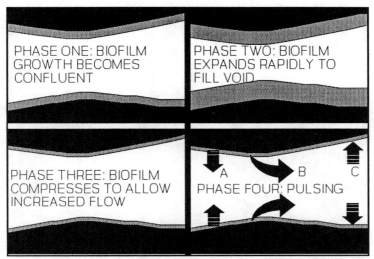

Figure 10. *First four phases of biofilm growth (shaded) in porous media (black, upper and lower edge). Phase four has three events: expansion (A), sloughing (B) and stabilization (C).*

may now have extended into the freely flowing water to cause radical increases in resistance to hydraulic conductivity. At the same time as the resistance is increasing, some of the polymeric material along with the incumbent bacteria will be sheared from the biofilm by the hydraulic forces imposed. Such material now becomes suspended particulates and forms "survival vehicles" through which the incumbent organisms may now move to colonize fresh econiches. In studies using laser driven particle counting, these suspended sheared particles commonly have diameters in the range of 16 to 64 microns (in-house laboratory data). In some cases, the variability in particle size can be narrow (+/-4 microns) around the mean particle size.

As the biofilm matures, another event which occurs is the stratification of the "slime" (Figure 11). The demands of the biofilm for oxygen (or nitrate) exceed the amounts being carried in the water flowing by. Consequently the

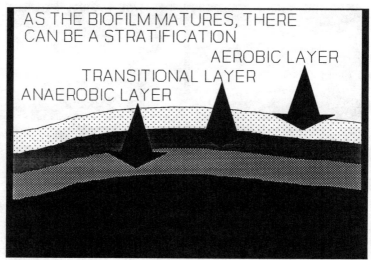

Figure 11. *A maturing biofilm will sometime stratify as it matures on a solid surface (black). Generally the deepest layer will be permanently anaerobic and be "protected" by the transitional layer.*

deeper layers become oxygen starved and hence turn anaerobic. This causes the microorganisms which flourish under environments free of oxygen to dominate. Two primary layers are therefore created. Aerobic organisms form a layer closest to the source of oxygen (i.e., the water) while the anaerobes congregate in the deepest strata where the oxygen never penetrates (i.e., permanently reductive). In between there would reside a fluctuating transitional zone which would sometimes be oxidative, and at other times be reductive.

BACTERIOLOGY OF PLUGGING

Biological impedance of hydraulic flows, whether in a water well, heat exchanger or cooling tower, involves a sequence of maturation within the biofilm prior to the plugging or other severe biofouling event. The types of microorganisms involved in such events vary considerably with the environmental conditions present. In water wells, the iron related bacterial group is commonly associated with plugging (i.e., significant loss in specific capacity due to a biofouling event involving biofilm formation)[21,22,45,51,65,72]. It should be noted that an alternate term to plugging is clogging. This latter term (clogging) is taken to mean any event dominated by chemical and/or physical factors which is causing a severe restriction or a total obstruction of flow from a water well.

3

The Diagnosis of Biofouling

Biofouling in a water well or in a groundwater environment is difficult to recognize for much of the evidence is indirectly obtained. Effects of the biological activity are observed rather than the *in situ* activity itself. The microbiological activities leading to a biofouling can involve a series of events involving primary colonization, biofilm formation, interlocking occlusion (which restricts hydraulic conductivity, flow) and stratified biofilm structures. These structures can generate a corrosivity potential due to the production of hydrogen sulfide and organic acids within the biofilms.

Biofouling occurs within the void spaces between the various relatively solid (and impenetrable) structures within the porous media. The growths which foul up these void spaces may take a number of forms besides a coherent film forming consistently over all of the surfaces (Figure 12). Three common forms of biofouling which are sometimes seen include thin threadlike processes which interconnect between different solid surfaces or are caused by biofilm shearing off from coherent surface growths into the void spaces moving in the direction of the water flow. Other growths can form as thick column-like structures which join between the various solid structures. This latter growth form can lead to the generation of an impermeable (bio) barrier which can totally restrict water flow and cause plugging.

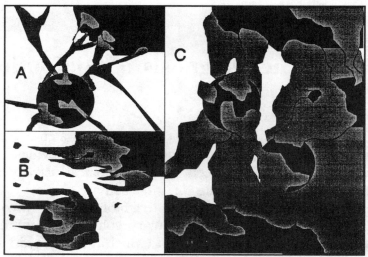

Figure 12. *Phase five involves a plugging of the porous medium (black) through the growth formations (shaded) which may be either threadlike (A), sloughing (B) or totally interconnected (C).*

Early diagnosis of biofouling involves direct determinations of the production capacity of the well, increases in the drawdown during pumping, upward shifts in the suspended particulate mass in the water as well as increases in the incumbent planktonic bacterial populations[51]. By ongoing monitoring of post-diluvial waters being pumped from a well which had been quiescent for a significant period of time, the level of biofouling can be projected using the data obtained based upon the sheared material observed in the water as it is subsequently pumped.

A series of biologically active zones (biozones) can be projected to occur around a water well where forms of biofouling are occurring. Predominantly, these effects can be categorized as (1) aerobic occlusive and (2) anaerobic corrosive. Various environmental factors can influence the rates of biofouling.

The life expectancy of a water well still cannot be accurately projected although some ranges have been published [48,72]. As a water well is developed and brought into service, various factors can influence the ability of that well to continue to produce a consistent quality of water in sustainable and reliable quantities. Production losses can be related on occasion to mechanical failure, the loss in the ability of the aquifer to deliver an adequate supply of water to the well[39]. This may be due to either too great a demand for groundwater being applied, inadequate replenishment or clogging/corrosion of the pump or column pipe. On some occasions, however, the malfunctioning water well may be more correctly diagnosed as suffering from a biofouling event.

BIOFOULING IN A WATER WELL

Events in which a biological component will cause a system or process to fail to achieve its expected performance criteria may be considered to involve biofouling. Such biofouling can be related to a biologically driven impediment in the process or a degeneration of one or more of the mechanical or chemical events considered essential to that process.

Classically, biofouling has been linked to many well documented corrosion events which have resulted in expensive remedial costs in the maintenance of water supplies. The most common examples of this are the activities of sulfate-reducing bacteria generating hydrogen sulfide which in turn leads to electro-chemical corrosivity of steels and concretes[42]. Less acknowledged is the role of microorganisms in the direct physical interference in a process through the generation of a biological mass which attaches to surface areas and changes various production and exchange criteria important to the process. Here, the attachment of the microorganisms to the surface areas first occurs and is followed by colonization and formation of a biofilm. Gradually this biofilm will thicken, forming copious polymer rich masses commonly referred to as slimes. The polymers composing much of the slime are very long thin chain-like molecules with an extreme capacity to retain water. Potential organic and inorganic nutrients and waste products are taken up, used or released from the incumbent microorganisms.

These slimes may grow sufficiently to obstruct and totally occlude (prevent) hydraulic conductivity (flows) through the porous and (less commonly) fractured structures surrounding the well. This would include the gravel pack around a well screen and could prevent water from entering the well (i.e., plugging).

OBSERVING BIOFOULING

There is inadequate amount of field data to determine in a precise manner the percentage of water wells which do suffer from biological impairment to flows leading to partial or total plugging events. Indeed, the methodologies to predict these occlusive phenomena are only now being developed. A sequence of events has become associated with the biofouling of a water well, but clearly one major component is the generation of an active biofilm within some region associated with the well which can subsequently cause either a bioimpairment to flow, a degeneration in water quality and/or a corrosion and plugging.

A recent study[51] of a community well at Armstrong in

western Ontario, Canada revealed that the flow rates were changing in a cyclic manner (Figure 13). The flow rate was controlled to a standard drawdown which meant that the production would be influenced directly by the inflow of water to the well. It was noted that where daily averages were taken of the flow rate there was a vacillation between improving flow (facilitative) and degenerating (intercedent) flow which had a harmonic of approximately eleven days. This cyclic change would indicate that the biofouling was in phase four (see Figure 10). While these cycles continued for over 300 days, there was a steady degeneration in the production from the well. The biofouling thus was having a critical effect on the capacity of the well to meet a designed production quota.

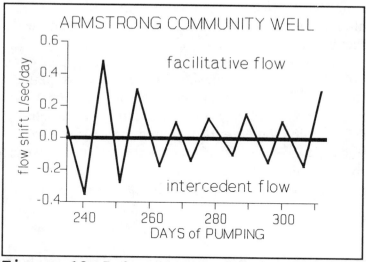

Figure 13. *Daily readings of flow at a standard drawdown revealed a pulsing change in flow rates (shift L/sec daily, vertical axis) sometimes improving (facilitative) or declining (intercedent).*

During the formation of such a biofilm, a number of

resulting events can be observed in the product water from the well. Critical amongst these changes is the concentration of particulates within the water. These may be associable with the biofouling event itself. As the biofilm grows within a groundwater system, it will entrap such metallic ions as iron and manganese within the polymeric structures. This may give different colours to the biofilm[50]. If the biofilm were to shear the resultant pigmented particulates (*pp*) would add pigments to the water. The common colour for these *pp* ranges from orange, red, brown through to black. Achromogenic particulates (non-pigmented, *ap*) may also shear from time to time away from the biofilm during the natural processes of maturation. Such sheared particles (*sp*) can be recovered from product water in amounts which will vary with the degree of shearing which is occurring from the biofilm into the passaging water at that particular time. In addition, water will normally carry a background of particulate material (*bp*) in relatively low density. It can therefore be extrapolated that there are three major components which could be present in the water as pigmented particulates (*pp*), achromogenic particulates (*ap*) which would form a variable part of the shearing particulates (*sp*), and in addition there are background particulates (*bp*). The total particulates (*tp*), which may be defined as all suspended particles which have a directly measurable diameter of greater than 0.5 microns, would be calculated as:

$$tp \quad = \quad bp \quad + \quad sp$$

where the *sp* is formed as the sum of *ap* and *pp*, both of which have originated from the shearing biofilm through the sloughing process and *bp* are the background particles already present in the water. These *bp* would be constantly present in the water regardless of pumping activity.

To assess the potential for the occurrence of a biofouling event within a water well, conditions have to be applied which would cause the biofilm to enter a shearing phase through an environmentally traumatic event. If biofouling is occurring, the trauma will cause an increase in the observable *sp* in the water.

TESTING FOR BIOFOULING IN A WATER WELL

When a producing water well has been kept quiescent for an extended period of time (e.g., seven days) there is a radical shift in the environmental conditions within the biofouled zone. There would consequently be a decreased availability of oxygen, nutrients and a loss of turbulent flows. These restrictors will cause stress in the biofilms composing the biofouling and one reaction to these stresses is for some of the biofilm to shear and become suspended sessile particulates. In this state, the microorganisms enter a phase involving an attempt to move and colonize a new econiche where the environment would be more favorable to the incumbent consortia. If pumping is again activated after this quiescent period, the pumping process will possess greater volumes of these suspended sessile particulates which would serve to magnify any biofouling event occurring.

It is clearly not very easy to directly observe the biofouling of a water well since the site of the infestation cannot be completely (or in some cases even partially) observed. Because of this restriction the evidence of biofouling has to be gathered through the observation of characteristic symptoms in the producing well (Figure 14). There are four major areas commonly recognized. These include: (1) reduced water quality (frequently due to some sloughing of the biofilms); (2) increased drawdown due to a "throttling" of the production capacity by occlusive biofilms around the well; (3) higher iron and/or manganese concentrations in the postdiluvial water due a combination of saturation of the biofilms with bioaccumulates and sloughing of accumulate-rich biofilms; and (4) very considerable increases in microbial loadings as a result of disrupting biofilms. Generally, all of these characteristics will occur together although the order in which they may be observed may vary from well to well.

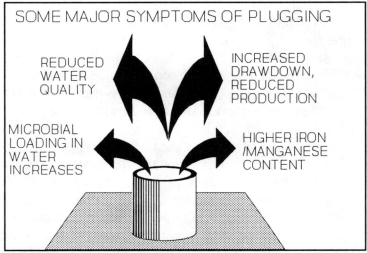

Figure 14. *Major symptoms of plugging of a water well (lower center) often occurs slowly with a variety of possible signals (captions to the arrows).*

The types of particulates which will be released from the well during the pumping will reflect the nature of the *pp* and *ap* which form the *sp* particulates. These particulates will give an indication of the types of microorganisms which are occupying at least the surface layers of the stressed biofilms. As pumping is continued, the particulates may reflect a combination of particulates arriving at the well from the more distantly located biozones around the well together with sheared particulates from deeper layers of the biofilms closer to the well. The projection of the precise position of biozones around the well within the zone of active water movement towards the well is, not surprisingly, difficult to determine.

There are three major events that have been observed when wells are pumped for the first time after a period of extended quiescence (stress to the incumbent microflora). In the first phase, where biofouling has occurred, all three types of particulates can be observed (i.e., *ap*, *pp*, *bp*). The water may have a distinct colored turbidity resulting from the dominating presence of the *pp* group or a milky type of cloudiness where the *ap* group is not overwhelmed by the *pp* particulates. The pigmentation in the *pp* group often originates from intense aerobic (oxidative) activity often associated with biozones close to the well water column and the gravel pack around the well screen being exposed to significant levels of oxygen. Oxidative processes particularly include the bioaccumulation of iron and manganese as these shift into the oxidative and relatively insoluble (ferric and manganic) states from the reduced (ferrous and manganous) more soluble forms respectively. Consequently, the *pp* around the well may be associated with these aerobic activities occurring closest to the well screen. Upon the initiation of pumping, any sheared pigmented particles (*pp*) will therefore tend to enter the pumped water quickly.

Non-pigmented biofilm may occur deeper out into the groundwater system beyond the direct influence of any oxygen associable with the well or vadose (unsaturated) zones above the groundwater. If there is a considerable concentration of sulfates in the water in the reductive zones around the well or more matured biofilms, there is a significant possibility of sulfate reducing bacteria (SRB) being active. These organisms cannot function unprotected in the presence of oxygen but do produce copious amounts of hydrogen sulfide which generate a blackening of the slime, black granulation (black gritty particles) and a rotten egg smell in the water. A black or grey *pp* may reflect a high or low presence of the SRB in an active state. If, however, there was a milky cloudiness dominated by *ap*, there would not have been the bioentrapment of oxidized iron and manganous salts and the slimes would have been achromogenic.

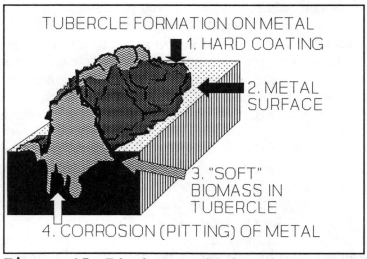

Figure 15. *Tubercles on metal surfaces often cover sites of corrosion. The "soft" biomass inside triggers corrosive processes on the underlying metal surface.*

One of the most covert forms of biofouling is possibly where a tubercle is formed on a metal surface (Figure 15). Here a complex structure is generated in which a hardened coating with a high iron (usually) content canopy forms over a soft biomass which interfaces directly with the metal surface. Microbially induced corrosion (MIC) may occur by a number of processes. The two most commonly understood are: electrolytic, involving hydrogen sulfide; or acidic, involving the generation of organic acids. The hardened outer coating provides protection to the incumbent biomass not only from predation but also from any chemical treatment applied (e.g., biocide) to control further corrosion.

As pumping is initiated, the greatest turbulence would occur closest to the well itself which would cause massive shearings at this point with a lesser effect as the distance of the influence of pumping continues to increase. It may therefore be expected that *pp* would be observed in the product water for extended periods of time reflecting the ongoing shearing of the biofilm further out from the well itself. These shearing episodes may tend to become more infrequent as the patterns of hydraulic flow stabilize with an ongoing pumping schedule throughout the well's "sphere of influence" in the surrounding groundwater. In confined systems, this adjustment of the hydraulic regime may be very rapid. This will result in a fast initial drawdown followed by a very slow decline to stabilization of the head. In such conditions a radical (initial) impact of the drawdown may be observed in a significant release of microorganisms from the fouling.

In general, shearing events will cause increases in the total suspended solids (TSS) which may range from 0.5 to >30 ppm TSS. Once the pumping has been extended sufficiently to stabilize the biofilm, the product water will contain only the background particulates (*bp*) which will usually amount to less than 0.5 ppm TSS.

MEASUREMENT OF PARTICULATE LOADINGS IN WATER

In the determination of a biofouling event, it is important to be able to differentiate and quantify the sizes of the *pp*, *ap*, *sp* and *bp* factions within the water. The pigmented particulates can be measured spectrophotometrically since the ferric and manganic salts, in particular, will absorb light and be easily measurable. Additionally, the water can have distinct colorations from yellow through orange to brown where very high concentrations of *pp* are observed. A simpler method to measure the *pp* is to entrap the particulates from a known volume of water on a membrane filter (MF) with a porosity of 0.45 microns. Here, through filtration the pigmented particulates will be entrapped on the porous surface of the MF. Once the MF has been air dried (to remove the reflective water films), the pigments may be recorded by inverse reflectivity using a reflectance spectrophotometer[22]. Here, the increase in absorbed light would indicate the concentration of *pp* through (inversely) a reduced amount of measurable reflected light.

The *sp* and *bp* particulates can be crudely measured using turbidometry or more precisely using a laser driven particle counter. These particle counters are able to size each individual particle as it is scanned and determine the composite volume and the number of particles of various predetermined sizes. From this data, the total volume of the suspended particles can be computed and registered as total suspended solids (TSS).

During the phase of pumping when the water is now dominated by a vacillating load of *sp*, the TSS values observed will fluctuate simultaneously. When the particulate loading stabilizes (usually at less than 0.5 ppm TSS) after the completion of "flushing", then only the *bp* will be observed in the product water.

EVALUATION OF MICROBIOLOGICAL COMPONENTS OF BIOFOULING

As water is drawn into the well from further out in the groundwater systems, the particulates recovered represent fractions from the incumbent biofilms now being subjected to a water flow which is overlayered with contaminating particulates still shearing from sites farther from the well screen. Confirmation of a biofilm event therefore requires that the particles recovered be confirmed as having a microbiological incumbency rather than being inanimate with a dead inorganic or organic composition. Confirmation of biological incumbency (viable entity presence) requires some level of biological testing at one of three potential levels: (1) **presence and absence (p/a)** to confirm whether organisms are indeed present in the particulate matrices; (2) at a **semi-quantitative level (s/q)** to determine the population of microorganisms occupying the particulate volume observed; or (3) **fully quantitative (q)** in which the density of the viable entities is determined together with the percentage incumbency of the particulate volume.

From experiences to date it would appear that the biofouling around a water well is not homogenous but involves a variety of structures. Easiest of these structure to observe is the iron rich biofouling which commonly occurs close to the well screen. The iron content makes these particular biomasses relatively easy to observe (Figure 16). These can take a variety of forms including: (1) tight regular spiculate in which protrusions extend roughly equally out into the aquifer in all directions; (2) irregular spiculate in which the protrusions tend to extend further out in the direction the water flow is coming from; (3) concentric detached occurs as cylinder of growth some distance from the well screen and usually relates to the conditions around the well being more oxidative; (4) dispersed, where there is a very "ragged" reduction-oxidation fringe and there is often relatively local recharges of oxygen and organic rich waters; and (5) tight concentric which is close to the well screen (not shown). These growths present only one form of the biofouling (high in iron and/or manganese bioaccumulates). Such growths may form into a distorted "top hat" shape

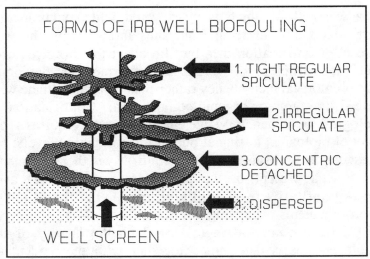

Figure 16. *IRB fouling around a well screen can take several forms. The diagram shows lateral sections of four of the common forms of iron-rich fouling which can occur.*

around a well or an irregular cylinder. A number of cylindrical zones of biological activity (biozones) can be postulated to occur around a water well, each of which would contain a different and distinct form of biological activity (Figure 17). These zones include: (i) the water column itself; (ii) the gravel pack or permeable media immediately surrounding the well screen; (iii) the zone beyond ii where there may be some extended but casual oxygen intrusion from the water well structure or operation; and (iv) the aquifer itself forming the source of the groundwater being drawn into the well by the effect of the pumping.

Biozone 1 is normally formed by the water column itself in the well bore. Because of this, there are some unique characteristics which will affect the microbial activity. These include a dominance of liquid water with a low surface area to volume ratio (for example, less than

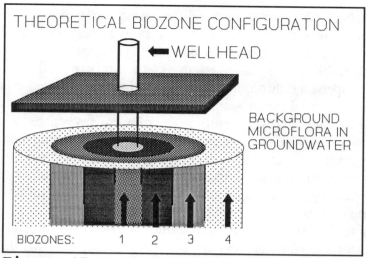

Figure 17. *Very often biofouling concentrates around a water well in a series of biozones. The innermost (1) is usually the most oxidative and the outermost (4) is the most reductive.*

0.1:1, cm^2:mL). During the passive non-pumping phase, a considerable stratification is likely to occur amongst the planktonic microorganisms present within the column. This will be at least partly due to a sedimentation of some of these viable particles towards the bottom of the well. The shearing potential from the limited biofilm occurring on the surface areas of the inner casing would be less likely to impart high levels of *pp*, *ap* and *sp* into the water during the initial pumping than resuspension of sedimented particulates as the turbulence inside the water column rapidly increases. Heavy biofouling in the pump itself also contributes to the microbial loading in this biozone. If the well is not sealed to prevent oxygen entering, there will be a significant diffusivity of oxygen into this biozone which will stimulate considerably higher levels of aerobic activity.

Biozone 2 is usually formed from the well screen or the intake area itself into the material immediately beyond the screen. This forms a very different econiche to biozone 1. Here, the surface area to volume ratio is very high ($>5{:}1$ cm^2:mL) with 20 to 50% porosity. This renders the zone vulnerable to reductions in hydraulic conductivity by bioimpedance through biofilm formation. There would be a high potential for oxygen intrusion, both from the well water column and through the movement of more oxygenated water down from the upper saturated zones around the outside of the well structures during pumping. The limited variety of types of surface areas and the relative availability of oxygen would suggest a high likelihood that any biofilm formation would be aerobic and could be dominated by the iron-related bacteria. At the initiation of pumping, the turbulence in this biozone 2 would be very considerable leading to a high level of biofilm shearing which would input *pp* and *ap* as *sp* into the water.

Biozones 3 and 4 form secondary econiches around the water well where the surface area to volume ratio would possibly be even higher, thus also allowing biological attachment and growth. In each of these subsequent biozones there would be reduced levels of turbulence. Oxygen intrusion may be expected to be much more restricted. Much of the oxygen arriving may be migrating vertically downward from the higher unsaturated zones rather than diffusing outwards from the well itself. There may be a higher proportion of microorganisms able to function both aerobically and anaerobically and hence able to continue to metabolize when this econiche enters a reductive phase (with the oxygen excluded) or oxidative phase. It can be therefore projected that there would be a decline in the size of the microbial population in this biozone where activities were restricted by the available nutrients. Background (biozone 5) would be outside of the concentric biozones. There would be the groundwater within the aquifer which is outside of either the direct zone of physico-chemical influence of the well. Here, the normal resident flora and such transient organisms would be present that may be passing through. These biozones would be expected to terminate within the hydraulic area of influence.

MICROBIOLOGY OF PLUGGING IN A WATER WELL

Biofouling plugging in a water well is caused by the biofilm occupying a sufficient volume within the interstitial spaces to cause a restriction in the flow of water into the well proper[39,48,51,72]. Such restrictions can function in four ways: (1) the biofilm formation within the interstitial spaces can directly restrict flow; (2) the shearing polymerics from the biofilm can cause increased resistance to hydraulic flow which is generated by the biofilm's surfaces sloughing; (3) gas may be evolved which remains entrapped to form a barrier; (4) additional waterborne accumulates entrap within the biofilm to cause radical increases in the volume of the biofilm which initiates an occlusion of the interstitial spaces. Combinations of these four mechanisms are capable of restricting hydraulic flow and, in doing so, cause the well to become partially or totally plugged.

In laboratory experiments[22,50], where the development of iron-related bacterial biofilms have been monitored, a number of phases have been observed. These phases reflect some of the potential activities which can occur within a biofouled water well. Maturation of the biofilms involved in a biofouling condition can create a plugging by passing through five phases.

Phase One, the biofilm becomes confluent in the biofouled zone.

Phase Two, radical increase in biofilm volume often reaching 50% or more of the total porosity of the medium. Parallel and equal to this is an increasing resistance to flow through the porous media.

Phase Three, a compression in the biofilm volume occurs which can reach as much as a 95% reduction from the phase one maximal volume attainment. In parallel, the resistance to flow becomes diminished and, on some occasions, the flow may actually be facilitated to speeds greater than that achieved in the control (unbiofouled) pristine porous medium. The size of the incumbent biofilm is now relatively small.

Phase Four, there is a gradual but slow increase in the volume of the incumbent biofilm. Elimination of any facilitated flow so that flow now declines below pristine value. This includes a continuing increase in the biofilm volume but this fluctuates in stepped harmony with changes (overall increases) in the resistance to flow.

The harmonic motion is triple phased. In sequence, the events include: (1) expansion of the biofilm; (2) shearing (sloughing) of the outer part of the biofilm; and (3) restabilization of the surface of the biofilm. These events have an impact upon the rate of water flow and on the water quality characteristics. Water flow is retarded in phases 1 and 2 and recovers in phase 3. Water quality tends to degenerate in phase 2 with the release of sheared particulates from the biofilm. In essence, the biofilm expands marginally while passing through the phase of expansion and then partially collapses during the ensuing instability. After the end of the third phase, the biofilm volume may have increased beyond that present at the beginning.

Phase Five, the biofilm has matured to occupy 60 to 80% of the interstitial spaces and reduces the effective porosity of the media. Interconnection of the various biofilm masses now will occur to cause a severe or total plugging with extreme resistances to hydraulic flows through the biofouled zones.

These plugging events can be summarized as: (i) a radical primary volume expansion; (ii) primary compression; (iii) secondary volume expansion; (iv) vacillative pre-occlusive state; and (v) total occlusion. The length of time which each of these phases occupy cannot be projected. In the laboratory, total plugging can be achieved in times which can be varied from two days to two years depending upon the environmental constraints applied.

FACTORS INFLUENCING MICROBIAL PLUGGING

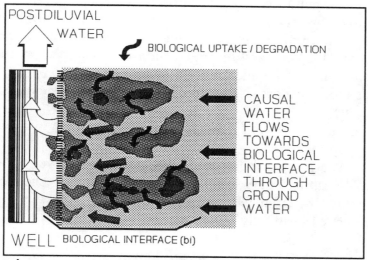

Figure 18. *Side section of a water well showing causal water (black arrows) interacting with the biological interfaces (shaded zones) whereby chemicals are removed by uptake/degradation (curved arrow).*

Causal and Postdiluvial Water

Plug formation in a water well is influenced by many physico-chemical factors[21,22]. During the process of bio-fouling in a saturated porous medium, there are a number of transient biological events that can be observed. For example, the causal water (*cw*) entering the well's zone of influence will interact with the biological events occurring around and within the water well. The water will therefore become subjected to the activities of these incumbent biological interfaces (*bi*). Water being pumped from the well has already been subjected to the influences of this *bi* within the biozones and can therefore be considered to be postdiluvial water (*pw*). Because of these interactions, the *cw* would become amended in its quality by these effects. The concentration of a given affected compound in the post-diluvial water (i.e., *pwc*) may therefore be significantly different from the causal water concentrations (*cwc*) as a result of uptake and/or degradation of each specific compound in the biological interface (*u/d:bi*). The amount of this biological activity could be theoretically determined as:

$$u/d{:}bi = cwc - pwc$$

There are a number of mechanisms involved in the biological interface leading to the removal of chemicals (Figure 18) from the *cw*. The size of the *u/d:bi* may be calculated factorially (*f:u/d:bi*):

$$f{:}u/d{:}bi = (u/d{:}bi)/cwc$$

The *f:u/d:bi* reflects a number of potential events which could occur within the active biozones around a water well. These events can include an initial bioaccumulation of chemical substances on or within the polymeric matrices of the biofilm itself. Once absorbed, the chemicals may now become subject to a passive temporary or permanent entrapment or become assimilated through various biosynthetic or biodegradative functions. Much of the iron and manganese so accumulated appears to be subjected to a passive entrapment while organic molecules are more likely to become sequentially subjected to a more complete assimilation through biodegradation or synthesis.

Iron and Manganese Biofouling

Iron and manganese are two major components in biofouling which have been subjected to considerable attention, partially because of their dominant observable presence in slimes[26,30,36,39,62,73]. When shearing occurs, the water generates a very pronounced pigmentation which can be readily seen. From field experiments to date, it would appear that iron tends to accumulate in biozones 1 and 2 close around a well. Manganese, on the other hand, tends to be more diffusively distributed through all of the active biozones. The sites of iron and manganese bioaccumulation appear to be in and around microbial cells and are relatively defined. The site of the accumulates can be: (1) extruded in an extracellular polymeric substance forming a twisting ribbon from the cell (i.e., *Gallionella*); (2) accumulated inside a defined sheath (tube of slime) formed around a group of microbial cells; (3) accumulated on the outside of the sheath which forms a tube around a number of microbial cells; (4) accumulated randomly in and around the polymeric slimes which encompass a number of microbial cells; or (5) is assimilated into the microbial cells directly.

Groundwater temperatures also have an influence on the biofouling. However, the variance in groundwater temperature is at a much lower order of magnitude than for surface waters. While surface waters can experience very significant diurnal shifts in temperature (e.g., $> +1 / -2°C$ per 24 hour period) and even more radical seasonal variations, particularly in the temperate zones. Groundwater, on the other hand, may exhibit seasonal variations ranging over as little as a 0.5 to 1°C range. Frequently, the fluctuations can be as high as 5°C seasonally for some shallower wells. The type of microbial colonization likely to occur within a groundwater system is therefore going to be dominated by microorganisms that are able to grow efficiently at ambient temperatures without experiencing major diurnal fluctuations. The most likely cause of temperature shifts would be the act of pumping, which would possibly move water towards the well from different depths where a temperature gradient may cause an elevation of the water temperature arriving at the well. Additionally, the heat generated by pumping may itself cause temperatures to rise.

With a narrow band of operating temperatures occurring in groundwater, there is more likely to be continuous support for either psychrotrophic (bacteria able to grow at below 15°C) or mesotrophic bacteria (able to grow within the temperature range of 15 to 45°C) throughout the year. The seasonal fluctuations so commonly reported for surface waters would therefore not be so applicable to any microbial activities occurring in groundwaters.

Fate of Organics in Groundwater

Organic materials arriving at biofouled sites around a water well may be in one of two forms. These are: (1) dissolved, and (2) suspended in a particulate mass moving through the water system. As the groundwater moves closer to the zone of influence that the well is generating in the surrounding groundwater, there may be some shifting of the dissolved organics into the mobile particulate state as assimilation occurs. Once the organic material arrives in the biozones surrounding the well, a number of interactions can occur[44,77]. These could relate to the absorption of the materials into the biofilm (i.e., bioaccumulation) or the degradation of the compounds from the aqueous phase (biodegradation). Organic materials entering the biozones resulting from the influences from the water well can be projected to enter the biological systems in a number of ways. This would include: (a) passive accumulation within the polymeric matrices of the biofilm, (b) active accumulation within the viable cells with subsequent degradation and (c) utilization for synthetic and energy-generating functions. As the water flows over the biofilm, it may be expected that some shearing of the biofilm will occur causing the releases of biofilm derived (sheared) particulates into the water phase. Thus, the passage of assimilable organics over an active biofilm may be expected to include these organics accumulated by the biofilm. Subsequent releases of amended and non-amended organics to the water can therefore occur in dissolved or sheared particulate states.

Microorganisms react in different ways to, and have various levels of dependency on, oxygen. These reactions may be categorized into a number of groups.

In the first group, oxygen is essential for the activity of the microorganisms (strict, or obligate, aerobes). Absence of oxygen prevents these organisms from metabolizing and reproducing. A subgroup is formed by those strict aerobes which is able to substitute nitrate for oxygen and continues to metabolize. This subgroup is known as the nitrate respirers. In groundwater systems these strictly aerobic organisms are likely to be found flourishing in oxygen rich zones (e.g., biozones close to and within a water well). Nitrate respirers may extend beyond the oxygenated zones around a well where nitrates are present. Another subgroup of aerobes is formed by those microorganisms which can function efficiently within only a limited (usually low ppm to high ppb) concentration range of oxygen. These are known as the microaerobes or microaerophiles.

The second group is the most adaptable in that these

organisms are able to perform in the presence (aerobically) or absence (anaerobically) of oxygen using different biochemical systems. These are known as facultative anaerobes. While they will grow in a wider variety of oxidative (oxygen present) and reductive (oxygen absent) environments, these organisms generally grow faster aerobically. It may be expected that biozones further away from the well may contain a substantial, or dominant, number of these facultative anaerobes.

In the last (third) group, microorganisms function in the absence of oxygen (strict, or obligate, anaerobes). Most of these organisms are very sensitive to the presence of oxygen, which is lethal to the cells. The reason for this is that, under conditions where oxygen is present, the cell produces peroxides which are lethal to the cell if not degraded. Anaerobes, by and large, do not have the ability to degrade peroxides and are consequently killed. Aerobic microbes produce enzymes (peroxidases, catalase) which break down the peroxides as they form. A few strictly anaerobic microorganisms do possess such enzymes and are able to continue to function in the presence of oxygen although they do not use oxygen in the metabolic processes. These are known as aerotolerant anaerobes. Generally, the third group are present in the deeper strata of biofilms formed on the reductive side of the fouling and are therefore more commonly found in biozones further out from the well.

Some microorganisms are oxygenic, that is, they are able to produce oxygen. This is achieved by the process of photosynthesis used by cyanobacteria, algae and the plant kingdom. Such oxygenic activities may be expected to occur unless light (solar or power-generated, bioluminescence) is present. Consequently, these oxygenic activities are not likely to be significant except where the water column is being directly illuminated. Around a water well, biological activities are likely to be relatable to the amount of oxygen present and the reduction-oxidation potential.

As the water approaches the well with an increasing velocity and level of turbulence, there can concurrently be a shifting in the reduction-oxidation state (redox) potential from a relatively reduced to a relatively oxidized state. The shifting in the redox potential can cause adjustments within the incumbent consortia in the biofilms. In the reduced state, the biofilm would tend to be dominated by the anaerobic organisms which do not necessarily require oxygen for growth. In oxidative conditions, at least the more exposed microbial components in the upper strata of the biofilm (nearer to the polymeric:free water interface) may be expected to be dominated by aerobes. In consequence, there would be a movement of organics into and out of the polymeric structures as the environmental conditions change. When the water arrives in the well water column through the screen, it could contain only some of the organic materials that were present in the causal water. These would appear in the dissolved form and/or as components within the sheared biofilm particulates. The level of organics therefore found in the postdiluvial water pumped from the well may be expected, in a well which has been subjected to significant biofouling, to be amended both in concentration and molecular forms with a mean reduction in the total concentration of organic carbon delivered (due to degradation).

Most of the microbial activities within a biofouled water well would, in fact, be attached to the surfaces presented by the pump, casing, well screen, gravel pack and the natural porous and fractured media occurring in the aquifer. Relatively little active microbial activity appears to occur in the planktonic (freely suspended) state. Most microbiological techniques for determining the occurrence of biofouling rely on an evaluation of the postdiluvial water for a determination of biological activity. However, if no shearing is occurring from the biofilm at the time of sampling, it can be extrapolated that a much lower population of microorganisms has been recovered. These "background" microorganisms may be the result of planktonic activities within the well water column itself (biozone 1) along with some resuspension of sedimented particulate masses from the base of the well. Because the biofilm may shear only periodically to release sessile bacteria from the biofilm to the planktonic or suspended particulate states, the absence of microbial indicators from the postdiluvial water does not necessarily indicate that the well is biologically pristine and has not suffered from biofouling. The ideal time to draw samples from a well in order to more correctly evaluate biofouling would be after a prolonged period of quiescence when the well was not pumped. During this period, biofilms may become traumatized by changes in environmental conditions. When pumping is again initiated, the turbulence generated will cause a magnifying effect on shearing and greater particulate loadings may be expected to occur, particularly from a well suffering from a significant biofouling.

Monitoring a biofouled well over this period of exaggerated shearing can be used to indicate, not only whether biofouling is indeed occurring, but also the extent to which the biofouling has infested the various zones around the well.

One mechanism for observing biofouling is to conduct a TV logging of the bore hole and observe the presence (or absence) of any biological "growths". The form of these growths is varied but the predominant forms (Figure 19) which can be seen are : (1) mucoid tubercles which often resemble cauliflowers or brussels sprouts in shape; (2) hard plate-like structures which are often tightly attached to the well screen itself; and (3) mucoid particles floating freely in the water which may have grown as sessile particulates or have been caused by sloughing from the mucoid tubercles. Not visible using the TV camera is: (4) covert biofouling which is occurring away from the well screen and so is not visible even through the slots of the screen. Such covert growths may be confirmed using the BAQC technology described later.

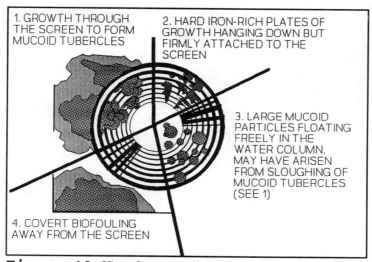

1. GROWTH THROUGH THE SCREEN TO FORM MUCOID TUBERCLES

2. HARD IRON-RICH PLATES OF GROWTH HANGING DOWN BUT FIRMLY ATTACHED TO THE SCREEN

3. LARGE MUCOID PARTICLES FLOATING FREELY IN THE WATER COLUMN. MAY HAVE ARISEN FROM SLOUGHING OF MUCOID TUBERCLES (SEE 1)

4. COVERT BIOFOULING AWAY FROM THE SCREEN

Figure 19. *View down a well at the screen level (concentric rings) depicting the different forms of biofouling (shaded zones) which may also extend out beyond the well.*

4

Direct Evidences of a Bacterial Problem

It has often been thought that a clear sparkling water sample is one that has to be free of bacterial contamination. This is not always the case since there can be as many as 400,000 bacteria in one milliliter of such a clean-looking water[2,13,23,32,47,54,59,75,82]. However, while the eyes may deceive on some occasions, other senses or time may be used to recognize a bacterial event through some obvious signals. Evidence that a bacterial problem may be occurring can take the form of a generating slime coating, floccular material or cloudiness which gradually intensifies in the water sample accompanied perhaps by changes to the taste, odor and color in the water. Each of these symptoms may relate to different problems being generated microbiologically.

Slimes are produced frequently where water may be flowing freely over a surface after passaging through a closed system such as a pipe. Slimes are usually considered to be gelatinous in nature and often glisten in reflected light. The most common colors for such slimes range from white, through grey to black, or through orange to red and dark brown. On occasions the red to brown-dominated slimes will shift to and from a black slime in response to changes in environmental factors.

These glistening slimes tend to be of a nongranular texture in nature and do not have a gritty texture. Exceptions to this rule are that some of the brown slimes which will generate considerable granular deposits of iron and manganese oxides and hydroxides which are very gritty in texture. Here the surface may form into a hardened crust. Such slimes bioconcentrate these oxides and hydroxides so that these growths may develop into forms of encrustations which will become hard and brittle as they expand. Within the encrustation or tubercle, microbial activity may continue and also initiate the processes leading to corrosion.

Regardless of the type of slime, all bear a common origin in that there was an initial growth of attached bacteria on a surface. Attachment is caused by the individual cells throwing out stringlike molecules (called polymers) which lock onto the target surface and anchor the cell to that surface (see SHOW Biofilm 1). Once anchored, the cells grow and multiply to colonize the surface with a coating (commonly referred to as a biofilm) generated through the cells rolling, leaping and spreading during reproduction and colonization. At the same time, more and more polymers are produced which bind the growing numbers of cells down to the surface being colonized. In addition to providing support, these polymers also take up and bind large volumes of (bound) water into the structures developed within the biofilm. A slime-like mat is thus created within which the bacteria can now develop complex associations (consortia) involving many different strains of bacteria.

VISIBLE MANIFESTATIONS

As the slime mat grows and matures, it goes through a complex life cycle involving rapid volume expansion, compression, a slow pulse-like secondary growth with periodic sloughing during which some of the slime may become suspended in and pass along with the flowing water. Slimes are therefore each complex in nature and ever changing in their makeup. Each major type of slime represents a different group of bacteria which, as a consortium, are dominating the biofilm. The following generalizations may be made for the slime types defined by color:

White or Clear slime - these slimes have not taken up any significant amounts of metallic salts which would cause pigmentation to occur. Concurrently, the bacteria are not generating any pigments of their own which could impart color to the slime. Care should be taken in the determination of color generated by bacteria because some may only be apparent in the ultraviolet light waveband. When present in high concentrations, these pigments may give a lightish green, yellow or blue color. Application of an ultraviolet light will then show a strong color reaction of the same type. Such a phenomenon occurs when the fluorescent pseudomonad bacteria are very prevalent in the slime. These types of slimes may be very rich in heterotrophic (organic nutrient-using) bacteria and can harbour a wide range of bacteria including coliforms where environmental conditions permit. Sulfur-oxidizing bacteria may also form these types of slimes particularly in a hydrogen sulfide-producing groundwater.

Grey slime - this type of slime forms an intermediate between a white (clear) slime and a black slime. Close inspection of the slime reveals that it will often consist of a white or clear slime within which there are intense black granules. These granules may range in size from smaller than a grain of sand to larger particles up to 7 or 8 mm in length. Such black deposits are most commonly metallic sulfides (usually dominated by iron and manganese) generated by interactions between the various metallic salts and hydrogen sulfide. This hydrogen sulfide may be generated by bacteria growing in the absence of oxygen (anaerobic) and reducing sulfur, sulfate (sulfate-reducing bacteria, SRB) or via protein degraders. Protecting these oxygen-hating SRBs is the clear slime in which oxygen-using (aerobic) bacteria grow in complex consortia.

Black slime - when the slime is intensely black throughout, this means that the bacterial flora is dominated by SRBs. In all probability this type of slime will be found in an environment depleted of oxygen but rich in the essential organic materials necessary to support these bacteria. Corrosivity is most commonly associated with the black slime due to initiation of electrolytic corrosion by the hydrogen sulfide generated by the microorganisms.

Orange,   Red, and Brown slimes - these slimes most commonly occur in environments that are relatively rich in dissolved oxygen (aerobic) and are exposed to transient amounts of iron and manganese in the water. A dominant group of bacteria in this type of slime are the Iron Related Bacteria (IRB). Here, the bacteria are unique in that they are able to take up iron and manganese in excessive quantities. The excess is deposited either around the cell, within the slime or special structures protecting or extending from the   cells. These are deposited as various oxides and/or hydroxides. Depending upon the precise combination of these oxides and hydroxides, the slime will generate an orange, red or brown slime as the bioaccumulated concentration   increases. These iron and manganese rich deposits appear to perform a number of roles from being protective (against predation and physical disruption) to being a nutrient reserve. These slimes form sites of intensive oxygen consumption which may frequently allow anaerobic bacteria to survive and grow within those parts of the slime where oxygen fails to penetrate. In consequence, these slimes may shift to a black phase when there is an oxygen depletion in the environment.

In examining the bacteriology associated with these slimes, two approaches can be pursued. These are: (1) determine the bacterial loading of the product water which has passed over the slime; and (2) remove some of the slime from the original site for direct microscopic evaluation.

Examination of the product water suffers from a severe disadvantage in that bacteria will slough off from the biofilm (slime) into the water in an irregular manner. Such sloughing events tend to occur in a random pattern from the slime to cause a very variable population to be recorded in the water over a period of sampling. Higher populations are likely to be observed when a water flow is suddenly generated by pumping over a quiescent slime rather than when there is a continuous stream of water moving across the slime layer in a hydraulically active manner. However, it is often much more convenient to sample this product water rather than to attempt to obtain samples directly from the slime, which may be growing at a relatively inaccessible site. For example, it would be difficult to obtain a sample from a brown IRB-rich slime situated at the outer edge of the gravel pack around a well screen. Indeed, in some circumstances, even with the assistance of closed circuit television (CCTV) or fiberoptic monitoring, it may still not be possible to even view the slime let alone obtain a sample.

In the practical world, the most common confirmation technique is to take a flowing or static water sample from as close as possible to the biofouled site. This would appear to be the less reliable method but essentially, it remains the only practical one to obtain such a sample.

Tuberculous growths are a variety of slime growth in

which the active bacterial growth occurs within a hardened salt and oxide-rich encrustation[30,45,50,83]. The outer shell takes on the form of a series of bulbous extensions stratified above and/or wound around each other. In color, these tubercles tend to be (externally) brown in color due to the high intrinsic concentrations of iron and manganese oxides and hydroxides. Frequently, fissures and cracks develop in the tuberculous shell so that bacterial slimes may become observable down these fissures. Corrosion is often associated with such tuberculation since anaerobiosis, a natural precursor to hydrogen sulfide development and electrolytic corrosion, can frequently occur deeper into the tubercle. IRB and SRB groups of bacteria are commonly associated with these tubercles. One theory (anode "snowball") is that the tubercle begins as a gas bubble (such as methane) entrapped under a biofilm. This biofilm subsequently hardens and the gas is utilized. The vacated gas vacuole is now colonized by incumbents of the biofilm.

Cloudiness - not only can slimes be seen growing in association with water-saturated porous media but microorganisms may also be seen occurring directly in the water causing cloudiness (an even haziness in the water), flocculant growths (visibly distinguishable buoyant particles often with ill-defined edges) and discoloration to the water. Each of these three characteristics can be used as evidence of a different type of microbial event occurring in the water. General interpretations of these observations are listed below.

General cloudiness - when the water is held up to a white light or a light is shone up through the water sample, the occurrence of a cloudy appearance may indicate that either there has been a heavy salt precipitation or there is an active dispersed microbial population in the water. In the latter case, the bacterial population which will cause this type of cloudiness may be in excess of one hundred thousand colony forming units (cfu) per mL.

Direct microscopic examination of the water can be used to reveal the cause of this cloudiness. Salts would appear as large crystalline structures in the water with clear hard edges while the bacteria would tend to appear as small indefinite shapes often oscillating in the water randomly by Brownian movement or moving in specific directions if the bacterium is capable of sustaining directional movement (motile) or as larger irregularly shaped structures. In either event, bacteriological examination of the water should be pursued.

Flocculent growths - some bacteria are able to grow in the water as large diffusive objects which appear to have a shape and are able to control their intrinsic density close to that of the water. This allows the bacteria to "float" in the water as large masses. On frequent occasions, these flocs may be colored from a light orange to a brown color due to the accumulation of iron and manganese salts. One group of bacteria found displaying this feature are the sheathed iron related bacteria. These bacteria exist for a part of their life cycle in a slime tube (called a sheath) and, on other occasions, the cells emerge from the sheath to produce a copious slime formation within the water.

Coloration of the water - either with or without cloudiness water will, under some conditions, pick up solubilized organic and/or inorganic material which colors the water. In groundwater, one of the most common sources of this color is the iron and manganese salts released after bioaccumulation within microbial growths around the well. Once the growth masses become saturated, the surplus salts may become solubilized or slough from the biomass and appear in the product water. Where these solubilized forms become present, there may not be a parallel increase in microbial numbers. Where the coloration event is primarily due to sloughing and is visible as a general cloudiness, an increase in bacterial numbers may be expected. In either event, the IRB bacterial group may be suspected to be involved at least to some extent. When there are suspended particulates in the water, it is possible to use this characteristic as a monitoring tool. This may be performed in a number of ways. These include the measurement of turbidity (for cloudiness) in the water, or enumeration and sizing of the particulates by a laser driven particle counter (which will also yield the total suspended solids volume, TSS). Where there are a large number of pigmented particulates a reflectrometric evaluation of the dried filtered particulates can also be undertaken.

ODOROUS SIGNALS

Different bacteria have been well known to produce some odors which can be considered distinctive for particular groups of bacteria. These odors can aid in the initial determination of the types of bacteria which may be dominating in the water. To detect these odors, the water needs to be collected within an odor-free container (such as a sterile and clean glass sampling jar). The water should occupy roughly 50% of the total volume of the container which should then be sealed. Vigorous shaking of the container will create an aerosol of water droplets which should harbour some of the odorous material. Clear the nasal passages with two or three deep breaths, loosen the seal (e.g., unscrew and lift the cap) on the water container and gently inhale. Any odoriferous chemicals are now more likely to be in the gaseous phase and hence be more detectable. Repeat to confirm the type of odor. If no odor can be detected, then warm the water to roughly 45°C and repeat the smell test. The higher the temperature increases, the greater is the potential for some of the more volatile odors to be detected.

There are a number of different odors which can be linked to the activity of different groups of bacteria. This can be very useful in reaching an early decision as to which bacteria groups are likely to be causing the problem and should be further investigated. The major odors which can generated by microbial activities in water are listed below.

Rotten egg smell - this odor is commonly generated by anaerobic bacteria functioning in oxygen-free environments and reducing either sulfates or sulfur to hydrogen sulfide (SRB) or breaking down the sulfur-containing amino acids in proteins to the same gas (by proteolysis). Water containing this bad smell may contain one or both of these groups of bacteria. Caution must be exercised when this gas is smelled since the nose can become quickly saturated with the hydrogen sulfide gas and no longer able to detect the presence of the gas. One conclusion which could be drawn from this type of odor event is that the water bearing this gas almost certainly originated from an oxygen-free zone where organic nutrients were present.

Fish smell - a very subtle odor which requires careful screening. The smell is similar to that commonly encountered around a fish retail outlet or processing plant. Many of the pseudomonad bacteria frequently generate these off-odors during periods of intensive growth. These pseudomonads are oxygen-requiring (aerobic) bacteria which are able to utilize different and often specific organic nutrients. Where pollution of water occurs, involving some specific compounds such as those associated with a gasoline release, these pseudomonads may become dominant. On some occasions, members of these pseudomonads can generate kerosene or oil-like odors.

Earthy smell - some microorganisms belonging to the streptomycetes (mold-like bacteria) and the cyanobacteria (algal-like bacteria) produce a group of odors which very much resembles the typical smells emanating from a healthy soil. These odors are grouped in a class called geosmins. When these odors are detected, there is a probability that there has been an aerobic growth (if originating from the streptomycetes) or a surface water-based algal growth (if cyanobacteria were involved).

Fecal (sewage) smell - raw sewerage and septic fecal wastes often generate a very typical odor. These odors can be generated by the enteric bacteria which commonly occur in fecal material. These bacteria are more commonly called the coliforms. The presence of such odors in water is a very strong indication that fecal pollution has occurred in the water and coliform testing should be performed as soon as possible since an acute hygienic hazard may exist.

Fresh vegetable smells - many of the green algae, diatoms and desmids will generate odors resembling different fresh or rotting vegetables such as lettuce and cucumbers. These odors would indicate that there may have been some recent entry of algal-rich waters where a bloom was just forming. Sour odors of rotting vegetation would indicate that the algal bloom was in an advanced stage of decay and high bacterial populations may be expected to be present in such waters.

Chemical smells - such as gasoline and solvents are more likely to originate from a pollution event rather than from the groundwater system where it could have been generated by microbial activity.   Where these types of odors are observed, a detailed chemical analysis for BTEX (benzene, toluene, ethyl-benzene and xylene) and/or hydrocarbons should be considered as a very high priority to minimize health risk.   When considered necessary, a gas mixture composition can be determined by a GC-MS analysis of a headspace sample.

5

Differentiation of Microbial Forms in Biofouling Events

Iron bacteria has been a common term traditionally used to identify the biological component in well plugging but these organisms can also be defined as iron related bacteria (IRB). These IRB participate in both the assimilation of iron and/or manganese into natural accumulates (iron oxidizing or precipitating bacteria) and the dissimilation of these elements away from natural accumulates (iron reducing bacteria) back into solution[18,36,39,66,72,78,83]. Many IRB are able to perform either major function. In the process of accumulation, the oxidized insoluble iron (ferric) and/or manganese (manganic) deposits are formed within either the cell or the surrounding extracellular polymeric matrices (glycocalyx) in such a way that the product oxides and hydroxides render visible orange to red to brown colors to the conglomerated growths (e.g., slimes).

These IRB bacteria can be subdivided into three major groups based upon the nature and site of the accumulated iron and manganese. These may be broadly grouped as:

1. Ribbon formers
2. Tube formers
3. Consortial Heterotrophic Incumbents (CHI)

3.A Aerobic heterotrophs
    i.   Unsaturated zones
    ii. Saturated zones

3.B Anaerobic heterotrophs
    i.   Corrosivity generators
    ii. Gas generators
    iii. General anaerobes

3.C Health risk heterotrophs
    i.   Aggressive transitory pathogens
    ii. Covert opportunistic pathogens
    iii. Marker organisms of health risk

3.D Heterotrophic iron mobilizers
    i.   Iron oxidizing bacteria
    ii. Iron reducing bacteria.

While groups 1 and 2 can be clearly differentiated (Figure 20), group 3 includes many bacterial genera which

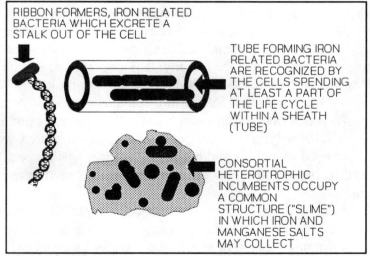

RIBBON FORMERS, IRON RELATED BACTERIA WHICH EXCRETE A STALK OUT OF THE CELL

TUBE FORMING IRON RELATED BACTERIA ARE RECOGNIZED BY THE CELLS SPENDING AT LEAST A PART OF THE LIFE CYCLE WITHIN A SHEATH (TUBE)

CONSORTIAL HETEROTROPHIC INCUMBENTS OCCUPY A COMMON STRUCTURE ("SLIME") IN WHICH IRON AND MANGANESE SALTS MAY COLLECT

Figure 20. *The major groups of iron related bacteria are separated by whether they produce a stalk (left, ribbon former), a tube (upper center, sheath former) or a consortium (lower center).*

function within more than one subgroup. Consequently, groups 1 and 2 will be described specifically, while the CHI group will be characterized by classical systematics.

RIBBON FORMERS, *Gallionella* (Group 1)

In the group 1, the ribbon formers, the iron and/or manganese oxides and hydroxides are accumulated within a spiralling ribbon of polymeric material which is excreted from the lateral (side) wall of each individual bacterial cell. Such ribbons can commonly exceed the length of the cell ten to fifty fold and include ten or more harmonic cycles in the length of the ribbon. Where these twisted, yellowish to brown ribbons are microscopically observed suspended in the product water from a water well, the dominant IRB in any associated biofouling is generally concluded to be *Gallionella* (see SHOW GAL).

Little is known of the life cycle for this bacterium and it has yet to be cultured in a pure culture. It is, however, clear that the bacteria producing these ribbon-like excrescences are Gram-negative rod-shaped or vibrioid cells. While occasionally the cells are seen to be still attached to the ribbon, often the ribbons do not exhibit an attached cell. This raises questions: why are most of the ribbons found in the water without the cells attached? and where can a bacterial cell grow within a heavily biofouled and turbulent environment and yet still produce such a ribbon?

One hypothesis[51] which would address these questions is that these cells have grown attached to the surface structures of the biofilm from which the ribbon-like stalks (false prosthecae) extend outwards into the zone of free hydraulic flow. The ribbon-like nature of the prosthecae would cause some of the transient flowing water to "spiral down" towards the attached cell in the manner of an Archimedian screw-like action (Figure 21). This would carry both dissolved oxygen and nutrients into the proximity of the cell. This would give the cells of the only genus in group 1 (*Gallionella*) an advantage over the cells that are incumbent in the biofilm itself.

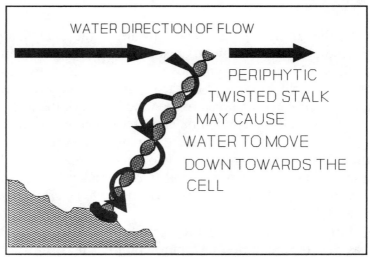

Figure 21. *Gallionella is a stalk (ribbon)-forming bacteria (black cell) which attaches to a surface (shaded lower left) and can cause water to flow down the twisted stalk towards the cell.*

As the prostheca (stalk) lengthens through growth and the accumulation of polymers and oxidized iron, so the shear forces imposed by the flowing water would increase until eventually the ribbon would break off. Once sheared, the ribbon would become an inanimate suspended particle in the water. By this means, fragmented ribbons would enter the water phase where they could easily be recognized and identified microscopically. Once the prostheca has sheared, the cell may now excrete a replacement stalk and continue to benefit from the ecological advantage this structure brings. The iron and/or manganese oxides and hydroxides would give a greater structural integrity to the prostheca and increase the survivability of the ribbon.

It is generally believed that *Gallionella* is able to gain at least some energy for ongoing maintenance and synthetic functions through the oxidation of these metallic ions from their reduced (ferrous) states. In general, this group is found in waters which have stressed nutrient loadings in the postdiluvial state. Total organic carbon (TOC) concentrations in waters bearing a dominance of *Gallionella* is usually relatively low, ranging often down to as little as <0.5 ppm and most commonly up to a maximum of 2.0 ppm TOC together with low concentrations of nitrogen and phosphorus.

TUBE FORMERS, *Crenothrix*, *Leptothrix* and *Sphaerotilus* (Group 2)

The second group of iron related bacteria also produce special structures which are formed as tubes of extracellular (outside of the cells) material. These structured tubes are called sheaths and the bacteria form filaments of cells which are able to move within the tube and even migrate out of the tube. Where this migration occurs, individual cells may become divorced from the filament and move away. This movement, where it occurs, is caused by flagella which may be tufted or a single polar flagellum. These organisms form a part Section 22, Sheathed Bacteria in the Manual (see SHOW BACT).

Iron and/or manganese oxides and hydroxides can become accumulated on and within the sheaths while the bacteria are able to function within the hollow central core of the tube-like sheath. Frequently, when scanning electron microscopy is conducted upon biofouled material, abandoned sheaths are commonly seen embedded in the surfaces of the biofilm. These bacteria tend to occur in waters with a relatively low organic carbon loading (<2 ppm total organic carbon).

When these sheathed bacteria grow, the sheaths may, for some genera, be attached to a surface by a holdfast. This allows the sheaths to retain the same position on a surface being subjected to hydraulic flows. Other sheathed bacteria either become enmeshed into the biofilm or the particulate growth suspended in the water (Figure 22).

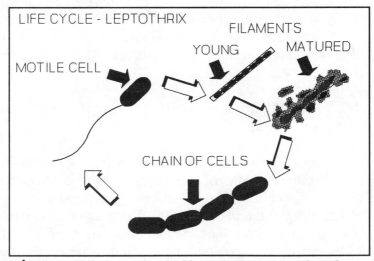

Figure 22. *Some sheathed bacteria (e.g., Leptothrix) have a complex life cycle growing from a young sheath (exposed) to a coated matured sheath which then releases chains of rods and motile cells.*

Cells are sometimes difficult to view through the sheath unless viewed by phase contrast microscopy. Wet-mount microscopic examination using 95% ethanol can improve the transparency of the sheaths and allow the incumbent cells and/or filaments to be more easily viewed.

The three dominant genera of sheathed bacteria of interest in groundwaters may be differentiated by the following dichotomous chart:

1. Sheaths encrusted with iron oxides over at least a part of the length? If yes, goto 2; if no, goto 6.
2. Individual cells clearly visible as cylinders or disks within the sheath. Tip of sheath clear while the base is coated with iron oxides (Figure 25). If yes, goto 3; if no, goto 4.
3. Genus *Crenothrix*.
4. Sheath heavily impregnated with iron and/or manganese oxides (Figure 25), sheaths not attached to surfaces by a holdfast. Cells in filamentous arrangement are difficult to view in the sheath? If yes, goto 5; if no, goto 8
5. Genus *Leptothrix*.
6. Sheaths commonly attached to surfaces by a holdfast, normally transparent with little or no iron oxide encrustation. Cells, where visible, are in filamentous form and the individual cells are not easily recognizable? If yes, goto 7; if no goto 8.
7. Genus *Sphaerotilus*
8. Is there a defined sheath which can be clearly seen resembling a tube in which individual or rows of cells can be seen with some branching of the sheath occurring? If yes, goto 9; if no, goto 10
9. Possible sheathed bacterium of another genus in Section 22.
10. May be a copious slime-producing bacterium which casually resembles a sheathed bacterium.

Sheath-forming bacteria are very common nuisance bacteria in water systems which have a relatively low nutrient loading. *Sphaerotilus* is an exception to this rule and is more commonly found in high organic environments. It is sometimes considered that, where there is a residual iron and or manganese in the postdiluvial water exceeding 0.2 and 0.01 mg/L, respectively, there is a greater probability of the sheathed bacteria belonging to *Leptothrix* or *Crenothrix*. Where there is a lower concentration, the dominant organism may be *Sphaerotilus*. These bacteria will tend to undergo a flocculative process when samples are taken from the well and subjected to very different environmental conditions.

Changes include a radical floc formation which may cause a visible "growth" to be generated which can cause system malfunctions within filters and treatment processes. Radical floc formation can be encouraged by exposing the water to a radical shift in the nutritional loading. Such floc formations can occur within hours of the water sample being taken. The floc formed is of a white to reddish brown color. Upon microscopic examination of a wet-mounted slide of the floc, it appears to be formed of a mixture of long very defined flexible tube-like structures which may have zones of darker brown encrustations around the tubes. In addition, there are poorly defined irregular masses of brown floc formed into conglomerates with the tubes.

These tube forming iron related bacteria are aerobic Gram-negative organisms and tend to be found in biozones 1 and 2 in and around the well water column in the well itself. These organisms may also be found growing in the capillary interface between the saturated and vadose zones above an aquifer. Where a well screen has been set close to this interface periodic releases of sheath-forming IRB may be expected to occur if there is a significant vertical downward movement of groundwater from that zone.

CONSORTIAL HETEROTROPHIC INCUMBENTS (Group 3, CHI)

Many microorganisms within groundwater systems do not function independently in separate niches but do cooperate with other species to form communities involving often many more than one species[2,6,13,20,47,50,55,64,75,79,82]. These community structures are referred to as consortia. For energy, most microorganisms in groundwater utilize the various organic fractions within the environment. These organic compounds may yield energy through various breakdown mechanisms to support the maintenance and growth of the incumbent microorganisms. Obtaining energy from such organic sources is referred to as heterotrophic function. The consortial heterotrophic incumbents (CHI) in groundwater therefore can be found growing within cooperative community structures generally within biofilms utilizing organic materials and other nutrients dissolved or suspended in the surrounding waters or within the glycocalyx structure itself.

It may appear surprising that such complex communities can form within environments where there would appear to be so little in the way of organic materials. The reason for this apparent anomaly is that the biofilm acts as a very aggressive accumulator of any passing organic materials[16]. Once retained within the structures of the biofilm, the organic material can be selectively used by the various incumbent microorganisms. There are a number of mechanisms involved in these degradative events. Indeed, the classification of microorganisms is, in part, based upon which groups of organics are or are not used[12,62]. Some of the major characterizations are given below.

Proteolytic, some microorganisms will degrade and use some or all proteins as a source of energy and amino acids. Proteins are usually in very low concentrations in groundwaters and so the most likely sources are other biological entities within or associated with the biofilm. Cannibalism, or the direct assimilation of cellular material from another viable entity, is thought to be a relatively frequent occurrence within a consortial community which can lead to changes in either the dominant species, reductions in the incumbency density and/or reductions in the variety of species present within the consortium.

Hydrocarbonoclastic, microorganisms are sometimes able to degrade different fractions of hydrocarbons. These organic compounds are very reduced organic entities predominantly composed of only carbon and hydrogen. Aerobic degradation (either molecular oxygen or nitrate appears to be required) does occur but at rates relatable to the complexity of the molecules (i.e., more complex, the slower the degradation). Anaerobic degradation may also occur but very slowly for even the more complex cyclic and aromatic hydrocarbons.

Saccharolytic, refers to the breakdown of various sugars. There are a wide variety of these sugars ranging from relatively simple forms such as glucose and sucrose to complex polymers such as starches. In general, these compounds are readily degraded by a wide variety of microorganisms aerobically. Glucose is one of the most universally utilized sugars. Where degradation is aerobic, the products are, to a large extent, either assimilated or released as carbon dioxide. Under anaerobic conditions, however, these compounds are only partially degraded, often with the releases of considerable quantities of organic acids. These acids are able to cause the pH in the environment to drop by as much as 4 pH units to create localized acidic regimes.

Cellulolytic, cellulose is one of the most commonly synthesized organic materials on the surface of this planet. The molecular structure renders the cellulosic polymers difficult to degrade and degradation is generally performed by relatively specialized organisms. Aerobic degradation is generally faster than under anaerobic conditions but is still a relatively slow process extending into weeks, months and even years.

Lipolytic, the basic subunits in lipids are glycerol and fatty acids falling into the general category of fats. Lipolytic microorganisms are able to selectively degrade specific lipids under either aerobic and/or anaerobic conditions.

Ligninolytic, woods contain a dominant amount of cellulose but do additionally include up to 30% of lignin. These lignins provide some of the structural integrity to the plant. The degradation of lignins has been found to be predominantly an aerobic function. It is performed efficiently by many filamentous fungi, in particular, the white rot fungi (WRF). These organisms tend to function where there is a considerable amount of oxygen such as in soils and the vadose (unsaturated) and capillary zones in aquifers. The WRF are thought to include some species also able to degrade polychlorinated biphenyls (PCBs) since there are some similarities in the molecular structure[50].

STRUCTURE OF CHI MICROBIAL COMMUNITIES

Before discussing in detail the various genera within the group 3 CHI organisms, it is important to comprehend the factors which cause these microorganisms to grow as complex consortial structures within the biofilms. These organisms can become stratified with the aerobic phase tending to be dominated by pseudomonads belonging to Section 4, family 1 of Bergey's Manual while the anaerobic strata may become dominated by Sections 7 (sulfate-reducing bacteria), 6 (anaerobic Gram-negative rods) and 5, families 1 and 2 (the enteric and vibrioid bacteria). The precise dominance sequence is a reflection of the nutrient and oxygen loadings of the causal water being delivered to the specific econiches.

Analyses

The amorphous nature of the sheared suspended particulates from such a biofouling renders microscopic examination frustrating. The more traditional bacteriological examinations using selective culture media such as R2A[72] and

Winogradsky Regina[18] media or simple biological activity and reaction tests (BART™, Droycon Bioconcepts Inc., Regina, Canada)[51] can be used to aid in the identification of the causative organisms.

Water samples being taken of pumped supplies from a well may not necessarily give an accurate indication of the microbial events occurring at the sites of biofouling. The water could also contain the intrinsic planktonic populations along with whatever sessile organisms that may be present in such sheared particulates as are suspended in the water. These occurrences are frequently random but a greater chance of observing these events can be achieved by entrapping the particulates in a moncell filtration system[49] or applying a pristine surface to the well water column onto which such organisms may now attach to form observable growths[72].

Application of Analyses

In practice, an understanding of the nature of a biofilm can be utilized in the management of particular intercedent events (i.e., through the generation of a biobarrier) such as in the control and bioremediation of a gasoline plume, removal of potential toxic heavy metals or aromatic hydrocarbons. These would be alternative events to the more appreciated roles performed when biofouling causes plugging and corrosion.

Environmental Factors

Many environmental factors will interact with the growth dynamics of a maturing biofilm[16]. These range through physical, chemical and biological factors[22]. Of the physical factors, the three pre-eminent constraints relate to temperature, pH and the redox potential where the site of activity is in a saturated medium. Temperature influences microbial growth in a number of ways ranging from the extremes where low temperatures cause an inhibition of the cellular metabolic processes, to high temperatures which may impact particularly on the protein constituents of the cell through thermal denaturation to cause the death of the cell. Between these two extremes lies a temperature range within which the cells can be metabolically active. Such a range will include a narrower band within which growth and reproduction can occur with maximal growth responses often being observed within relatively narrow temperature ranges (e.g., 10°C). For temperate region groundwaters in shallower aquifers of up to 300 meters depth, the normal temperatures experienced without geothermal influences could range within a band from 1° to 25°C. This range is commonly associated with bacterial activities dominated by the lower mesotrophs (>15° to <45°C with optima at <35°C), eurythermal psychrotrophs (grows both above and below 15°C, and stenothermal psychrotrophs (grows at <15°C only).

Temperature will therefore influence the makeup of an incumbent consortium within a biofilm. For example, a water with a temperature of 10°C (+/- 2°C) may be expected to include eurythermal and stenothermal psychrotrophs but no low mesotrophs except perhaps around the heat generating pump motors. This does not preclude the potential for mesotrophs surviving prolonged periods under these conditions. Shifting groundwater temperatures may be expected to cause changes in the dominant microbial strains within the biofilm.

The polymeric structures in the biofilm will tend to impart a buffering activity which reduces the influence of any pH shifts in the water upon the activities of the incumbent organisms. In general, the maximum species divergence in incumbent flora should occur when the pH of the passaging water is between 6.5 and 9.0. Where the pH falls periodically to below this pH range, the biofilms have some capacity to buffer these effects. Additionally, the pH within a biofilm may become stratified with higher neutral or slightly alkaline conditions occurring in the upper aerobic zones while the anaerobic zones may become more acidic and support, as a result, a narrower spectrum of microbial activities. In severely acidic conditions (e.g., pH <2.5) the spectrum may become so narrow that it includes only one or two different genera of microorganisms. For example, many of the sulfur-oxidizing bacteria belonging to the *Thiobacillus* species are able to function at very low pH values during the oxidation of sulfur and reduced sulfur compounds leading to the exclusion of other species.

In ecological investigations at sites of magnified biofouling in saturated porous media, it has often been noted that these sites tend to be concentrated in the transitional zones between reductive and oxidative regimes[15]. Here, the redox values shift from a negative to a positive Eh value (e.g., from -50 to +150 mV).

Sequential dominance of the various CHI bacteria have been observed occurring along these redox gradients. Such sequential events can be manipulated through a shifting and/or enlarging of the redox transitional zone. This can be achieved within an aquifer by recharging the edge of the reduction zone with oxygenated water to force the biofouling outwards (and away from the producing water well). This extension of the area occupied by the biofouling away from the well causes the bioaccumulation associated with the microbial activities to take place more distantly from the well. The postdiluvial water may consequently be of a higher quality since there would have been a slower passaging of the causal water through the biozones which are now further away from the turbulent zones of influence created by the pumping action within the well.

Nutrient Factors

Chemical factors can also influence the spectrum of CHI activities associated with these biofouling events. These factors can be summarized into nutritionally supportive or inhibitory groupings. The nutritionally supportive group would include compounds able to be utilized by the microbial cells for catabolic (energy-yielding) and synthetic functions essential to the continued survival, growth and dissemination of the species. Inhibitory groupings may be defined as any chemical which will, at a given concentration, directly interfere with any of the supportive functions of a given species in such a way as to at least minimally retard specific competitive functions, or maximally cause the death of that species group.

In the processes involved in natural competition between microorganisms, the supportive chemicals for one strain may, in fact, be inhibitory to another species. It may therefore be commonly expected that, where an inhibitory compound is introduced into an environment, there will be a radical shift in the component species within the impacted flora culminating in a tolerance of and, secondarily, a direct utilization of the compound by the surviving species. Additionally, the polymeric matrices of the biofilm will form a protective barrier through which such inhibitors would have to pass prior to directly impacting on the incumbent microbial cells. Such matrices can often form a repository for toxic bioaccumulates away from sites where there could be direct inhibition of the activities of the incumbent cells.

Nutritionally supportive chemicals are essential to the ongoing survival and growth of any given species. The nutrient elements essential to growth processes can be determined by an examination of the ratio of these elements in the microbial cell. A typical ratio for the elements $C:O:N:H:P:S:K:Na:Ca:Mg:Cl:Fe$ can be expressed gravimetrically as percentages of the dried weight (for *Escherichia coli*) as $50:20:14:8:3:1:1:1:0.5:0.5:0.5:0.2$, respectively.

Particularly critical to the growth of microorganisms is the $C:N:P$ (idealized above at $50:14:3$) ratio since these are the three macronutrients that are most frequently observed to be the critical controlling factors in microbial growth. If there was a 100% efficiency in the utilization of these elements, these idealized ratios in the nutrient feed could be expected to be optimized around N(=1) as the C:N:P ratio of 3.6: 1: 0.21 respectively. In reality, for the heterotrophic microorganisms, the organic carbon consumed far exceeds the maintenance requirement due to the heavy catabolic demands to create energy. The totally oxidized carbon is often venting as carbon dioxide or, when totally reduced, it can be vented as methane. The net effect of catabolism is to shift the C:N ratio downwards.

An additional diversion of carbon would be in the synthesis of ECPS formation outside of the cell. The optimal ratios for the efficient growth of heterotrophs has not been established for the C:N ratio due to the natural variations that occur between the catabolic and synthetic functions. However, for the N:P ratio, the bulk of these elements are retained within the cells so that it has generally been considered optimal within the range of 4:1 to 8:1. Excessive levels of phosphorus (i.e., ratio of <4:1) cause possible buildups in the reserves of stored polyphosphates or a greater probability of the flora shifting to those species able to undertake dinitrogen fixation (a mechanism in which the fixated nitrogen corrects the N:P ratio).

In generating a comprehension of the impact of organic carbon on the growth of sessile or planktonic microorganisms, concentrations of 1.0 ppm (mg/L) of organic carbon would appear at first to be insignificant (i.e., too low to support growth). However, where a saturated environment contained one million viable units of bacteria per mL and the dried weight of each cell was 2×10^{-13} g and a 50% gravimetric carbon composition, the combined net weight of carbon in each of these viable units would be 1×10^{-7} g. If these organisms were in an environment with a total organic carbon (TOC) of 1×10^{-6} g/mL of total volume, the ratio of cellular:free organic carbon would be 1:9. If one thousand bacteria were present per mL as viable units, then the cellular:free organic carbon ratio would shift to 1:9,000. It can therefore be projected that the critical dissolved organic carbon which could influence microbial activities in water should be considered to lie within the ppb range rather than the traditionally accepted mg/L (ppm) range. In developing a system to generate the bioaccumulation and/or biodegradation of specific nuisance chemical compounds in the environment, it becomes important to set the C:N:P ratios in a manner beneficial to the required activity. If the target compound is organic, it may contribute to the carbon ratio as a result of degradation and/or assimilation. A typical C:N:P ratio to be established to maximize microbial activity would range from 100: to 500 : 1 : 0.25, depending upon the amount of carbon that may become incorporated into the ECPS, the rate of biological activity on the targeted compound and the availability of alternate carbon substrates.

CHI GENERIC GROUPINGS OF SIGNIFICANCE IN GROUNDWATERS

The understanding of the bacterial groups which play a part on occasions in groundwater systems is complex[2,4,6,69,82]. It is complicated by the fact that there are two major groupings: (1) **Indigenous Bacteria** that occur naturally in groundwater systems. There is also a growing body of evidence that there are stratified within the earth's crust a series of distinctive types of microbial activity (intrinsic); and (2) **Opportunistic Bacteria** that naturally occur in another part of the biosphere but can, under suitable conditions, invade and compete within groundwater systems (extrinsic). This latter group can include possible pathogens which may be subdivided into two groups on the basis of their relative pathogenicity. **Pathogenic Bacteria** include those bacterial groups which are able to routinely cause, where the numbers of invading cells are adequate, clinical symptoms of the zdisease in an infected host. Examples of pathogenic bacteria include those causing cholera, typhoid and dysentery, all of which can be waterborne. **Nosocomial Pathogens** (opportunistic) is the name given to those bacteria which occur in the natural environment as a normal part of the eco-system but which can, upon entering an immunologically or physically weakened host, cause clinical symptoms of infection. These nosocomial pathogens may not necessarily show consistent clinical symptoms of the disease and so have traditionally been more difficult to diagnose. Most attention has been paid to hospital-induced nosocomial infections (where a patient has become infected with a nosocomial pathogen while resident in the hospital)[32]. Community- induced nosocomial infections also occur but are much more difficult to diagnose due to the dispersed nature of the infected hosts as compared to a hospital.

While there is a natural interest in the health (hygiene) risks[10] that may be associated with groundwaters[26], there are also risks to the "health" of the groundwater system itself[54]. These risks would relate to the influence that microbial growths and activities may have on the productivity of aquifers and the transmissivity of those waters within, between and into such aquifers. An aquifer in which microbial intrusions had in some way impaired productivity would be reasonably deemed to be a less "healthy" aquifer. Microorganisms may be subdivided into two major groups.

(1) **Occlusive Microorganisms** include those microbes which will grow within the porous structures of the aquifers and impede the transmissivity of the groundwater through the

system (and towards a producing water well). These organisms usually form in biofilms (slime) growing over surfaces and gradually forming a coherent growth which may totally block water movement (biological barrier or plug)[50].

(2) **Nuisance Microorganisms** interfere with the quality (rather than the quantity) of the groundwater passaging through and/or being delivered from the aquifer. These interferences may take the form of noxious products of microbiological activity such as detectable tastes and odors, cloudiness and color generation. These effects may be associable with either the products of growth (chemicals or dispersed cells) or the biochemical modification of the chemistry of the water due to such events as enzymic activity, shifts in the reduction-oxidation state, temporary microbiological entrapment with phased releases of the bioconcentrated materials and the releases of odorous chemicals.

Both occlusive and nuisance microorganisms tend to predominantly grow in consortia. Such growths involve the growth of a variety of microorganisms within a common biofilm or "slime". These forms of growth tend to focus at the interface between an oxidative (oxygen present) and reductive (oxygen absent) regime. The REDuction - OXidative status can be measured or calculated and expressed as the REDOX potential (Eh value) and the dominant microbial activity appears to occur at the REDOX fringe. One reason for this focus is that the oxygen (for aerobic growth) comes along a gradient from the oxidative state while the nutrients diffuse along the gradient from the reductive side of the gradient. In a highly oxidative state, there would tend to be shifts in the dominant microbial flora in favor of the molds particularly where the porous medium is not permanently saturated. In this event, taste and odor problems may become more pronounced with earthy and musty odors often dominating.

Under some conditions where radical contamination (pollution) of a groundwater system with an organic chemical has occurred there is likely to be phased microbial response to this event. The phases could include entrapment into the biofilm, partial or complete degradation, suspension and mobilization of any undegraded or partially degraded chemical products incumbent in any detaching biofilm and radical dispersion of any volatile or gaseous insoluble products. Three major events may therefore be associated with these events: **bioentrapment**, **biodegradation** and **biomobilization**.

Bioentrapment, the act of a chemical or group of chemicals being taken up (and potentially bioaccumulating) in a biologically derived structure (e.g., biofilm, suspended sessile particulate).

Biodegradation, the destruction of at least a part of a molecular structure by a biologically derived chemical process (e.g., enzymic action). The products of this degradation are normally one or more smaller molecules (e.g., benzene, lactic acid, carbon dioxide, hydrogen, nitrogen).

Biomobilization, an event where a bioaccumulated compound is released back into the water flow through either a sloughing of the biofilm containing the accumulate or the direct release of the compound from the biofilm into the water.

These events would be in addition to the restriction of contaminant mobility by occlusive (bio)barriers. In general terms, the microorganisms associated with such events are often considered to be **Biodegradative Microorganisms**. This would be true where there is pollution of a groundwater system by a specific group of organic chemicals. Here, it can be expected that the range of microorganisms focusing in the biodegradation zone will contain fewer types in the consortium. These will be restricted to those strains able to "cooperatively" degrade the compounds with a maximum of efficiency given the environmental conditions presented. The density of the incumbent degraders in the biofilm may reflect the level of aggressivity with which the targeted compounds are being degraded. In cases where some of the pollutant chemical is now becoming dispersed with shearing fractions of the biofilm, the level of degradation may be reflected in the density of cells recoverable from the suspended particulate (sheared biofilm) material.

It has to be remembered that when a water sample is being examined for microbial presences, that water sample will contain those components which either naturally occur in the water or have entered the water in sheared material from the biofilms attached to surfaces over which the water has moved.

Absence of microorganisms from a particular water sample cannot be construed to mean that there would be no microorganisms present within the groundwater system itself. Much of the observations on groundwater microbiology do relate to the microorganisms recovered from postdiluvial water samples and may not reflect the total range of organisms which may be actively associated with the groundwater system itself. There are, however, a number of major groups of consortial heterotrophic microorganisms which have been associated with various major events in groundwater systems. Each will be described briefly below but it should be noted that references to sections and families (bracketed) refer to the current classification of bacteria as described in Bergey's Manual (9th edition). Additionally, it should be recognized that there are two major groups of bacteria differentiated by a staining technique known as the gRAM (or Gram) stain. RAM refers to the ability of the stain to differentiate (R)eaction, (A)rrangement and (M)orphological characteristics of the bacteria. Particularly important in the gRAM reaction is the reaction differentation which may be negative (gRAM-negative, G-) or positive (gRAM-positive, G+)[62].

Pseudomonads, Soo"do-mo'nads (Section 4)

These bacteria are all gRAM-negative relatively primitive rods and cocci. All are aerobic and need oxygen (some can use nitrate as an alternative) in order to grow. Many are able to break down specific organic materials into inorganic substances (mineralization process), particularly under saturated oxidative conditions. These bacteria often grow in consortial biofilms in association with other microbial groups. One common example is the association of pseudomonads with sulfate-reducing bacteria in corrosive biofouling (MIC). Many pseudomonads are able to function very efficiently at low temperatures. Generally, these organisms are able to grow well at temperatures of less than 15°C and are called psychrotrophs. In oxidative groundwaters having temperatures of between 2 and 12°C, the pseudomonads may be found to dominate, particularly where there is a significant organic content (>0.5 ppm TOC) originating from a limited range of pollutants. In the case of liquid hydrocarbon plumes generating in groundwater systems, the range of pseudomonads that are recovered can become very restricted (i.e., 2 to 3 strains)[20]. Where a shallow monitoring well becomes biofouled under oxidative conditions, pseudomonads may well dominate the fouled zone and cause significant reductions in the recoverable pollutant withdrawn by sampling the well due to the preferential bioentrapment and biodegradation in the fouled zone around the monitoring well.

Some of the *Pseudomonas* species are pathogenic. Of

these, *P. aeruginosa* is the most serious and can infect people with low resistance to cause urinary tract and lung infections, and also invade burn areas. Two unusual families of pseudomonads may also commonly occur in groundwaters. One group generally referred to are the methylotrophs. These are able to oxidize methane gas aerobically. Methane is commonly found in anaerobic soils, sediments and aquifers and is generated (methanogenesis) where there is a reductive organic decomposition occurring. The second unusual family is formed by the pseudomonads able to grow in oxidative environments where the salt concentration reaches greater than 15%. These form the halotrophic pseudomonads.

Enteric Bacteria, En"ter-ic (Section 5, family 1)

These bacteria are also gRAM-negative rods but are able to grow under both oxidative and reductive conditions (facultative anaerobes). Most are fermentative degraders producing organic acidic products and often copious amounts of gas (usually CO_2 and H_2). The name "enteric" focuses on the fact that several of the genera inhabit the gastroenteric tract of warm-blooded animals and are also evacuated in significant numbers in fecal material. It is for this reason that the enteric bacteria have been used as indicator organisms for fecal contamination (and hygiene risk)[31,32]. As the major inhabitant of the human colon, *Escherichia coli* has been selected as a significant indicator of fecal contamination. *E. coli* can be pathogenic, causing gastroenteritis or urinary tract infections. However, several other enteric genera include major pathogens causing such diseases in humans as: typhoid and gastroenteritis (*Salmonella*), dysentery (*Shigella*), pneumonia (*Klebsiella*) and the plague (*Yersinia*). Amongst the vibrioid bacteria in the related family 2 is another major gastroenteric pathogen (*Vibrio cholerae*), the causative agent of cholera. Common tests for hygiene risk in waters involves an evaluation for the presence of **Coliform bacteria**[14]. These bacteria are defined as nonsporing gRAM-negative bacilli which are able to ferment the sugar lactose with both acid and gas being produced in determinable amounts. It is interesting to note that the sugar lactose is not commonly found throughout the environment but appears to be synthesized the most commonly in mammalian milk. Mammals when suckling during infancy therefore take in large quantities of lactose in the milk which biases the microflora in the gastroenteric tract to those bacteria that can anaerobically ferment lactose (e.g., coliforms). These coliform bacteria also have to be able to resist the inhibitory effects of bile salts (excreted into the gastroenteric tract) and be able to grow at warm-blooded temperatures (e.g., 35°C). Because of these restrictors, coliforms may be selectively grown in culture conditions where other contaminant bacteria would be suppressed. The restricting factors are the use of lactose as the major energy source; application of bile salts to selectively restrict competition to those bacteria normally found growing in the gastroenteric tract; and the use of an above environmental norm temperature for growing these coliforms (i.e., 35°C). Acid products are easily determined by color shifts in pH indicators and the gas may be entrapped for direct or indirect observation[10].

Confirmatory tests need to be performed if the presence of *Escherichia coli* is to be confirmed. The need to confirm that *E. coli* is present is important because some of the other coliform group are able to not only survive but also grow within the natural environment. Such microorganisms in a water system would therefore cause a positive coliform test.

That test, however, remains presumptive until the presence of *E. coli* is confirmed. The most common genera causing these types of interferences are *Enterobacter* and *Klebsiella*.

Sulfur Bacteria, Sulfate-Reducing Bacteria (Section 7)

Sulfur bacteria is an unusual name since it is used to describe two very distinct groups of bacteria. Commonly, the term is used to refer to the sulfate-reducing bacteria (SRB) which are associated with MIC. The term is, however, also used to describe the bacteria in whose activities sulfur plays a major role (e.g., producing elemental sulfur, sulfuric acid). These activities will be addressed separately. SRB are serious nuisance organisms in water since they can cause severe taste and odor problems and initiate corrosion. These bacteria are called the sulfur bacteria because they reduce large quantities of sulfates to generate hydrogen sulfide (H_2S) gas as they grow[42]. These bacteria are referred to as the sulfate-reducing bacteria (Section 7) and are often known by the initials SRB. The problems generated by H_2S are: (1) it smells like "rotten eggs", (2) it initiates corrosive processes, and (3) the gas can react with dissolved metals such as iron to generate black sulfide deposits.

For the plant operator each of these effects can become a serious nuisance. For example, the sudden appearance of the smell of "rotten eggs" in a water. Usually such events mean that somewhere upstream there has been a major aerobic biofouling which has removed the oxygen out of the water and allowed these bacteria to dominate. Rotten-egg smells are not created by just the SRB group but can also be sometimes generated by other bacteria such as many of the coliforms when suitable conditions occur. Hydrogen sulfide may be found in waters where oxygen is absent and there are sufficient amounts of dissolved organic materials present.

Forcing oxygen (as air) into the water can stress these SRB microorganisms since it is toxic to their activities. Commonly, the SRB protect themselves by cohabiting slimes and tubercles with other bacteria which are slime-forming. These rotten-egg smells will occur more commonly when a water system (or water well) is not used for a period of time. In these cases, the oxygen in the water is used up by various other bacteria cohabiting the biofilm. Once the oxygen is controlled by these other cohabiting members of the consortium, the growth and activities of the SRB group can become rampant. Sometimes this is accompanied by the intense production of the rotten-egg (hydrogen sulfide) smell.

The SRB group may establish itself attached to a solid surface within a biological slime or tubercle (a "bubble-like" structure composed of hardened iron-rich plates overlayering an active slime formation inside). Once established, these bacteria will begin to generate the hydrogen sulfide gas. This gas can trigger a complex electrolytic corrosion process on some metallic surfaces. Such corrosion begins with pitting and terminates with perforation of the supporting structures (e.g., metal pipe wall) and system failure. At the same time, various slime-forming bacteria cohabiting the site may magnify the problem by generating various organic acids which can also be corrosive. Once established, the corrosion is difficult to control since the slimes or tubercles "buffer" the effects of any chemical treatments such as chlorination, acidization or the use of cleaning agents. This means that higher dosages and longer exposure times need to be applied to try to control such corrosion events. Because the SRB organisms cause corrosion while growing on a surface, tests on the water itself may be negative. Positive tests will occur if some of these bacteria are present in the water while moving from one corrosion site in an attempt to colonize another site.

Where there is a significant amount of iron, manganese or other metallic materials in the water, hydrogen sulfide can react with these compounds to form metallic sulfides. Many of these chemical compounds are black in color and can cause the slime to become black in appearance. When these blackened slimes break up, the water may contain thread-like strings of black slime-like material. On many occasions, these black growths are not accompanied by any rotten-egg smells since the H_2S gas has been converted into the black sulfides. On some occasions, these slimes flow down walls, across surfaces and may even change in color through to browns, reds, yellows and greys as the oxygen in the air stimulates aerobic bacterial activity in the slime.

Identifying an SRB problem is easier to achieve by recognizing the rotten-egg smell, finding tubercles and corrosion inside metallic equipment and\or the black slimes rather than performing bacteriological tests on the water that has flowed past the site. The reason for this is that the SRB bacteria grow in "protected" places often surrounded by other types of bacteria which may mask their presence. Microscopic examination is also made more difficult because of interferences caused by these growths. There are some cultural systems that can be used to confirm the presence of SRB but most rely upon these bacteria producing black sulfides which become visible. This may be seen as a deposit in a suitable liquid culture medium or as distinctive black growths in and/or around colonies containing SRB growing on agar (gel) media in the absence of any traces of oxygen. The lag time before these "blackenings" occur can indicate the aggressivity of the SRB group. A rapid (e.g., two-day delay) would indicate a much more aggressive population than in a case where there was a considerable delay (e.g., eight days).

Risks from an active SRB infestation are multiple in that there can be corrosion, severe taste, odor and colored-water problems, losses or total failures in process efficiencies and increasing consumer complaints. Unfortunately, the ability of these bacteria to grow in places where they are protected by either copious "overburdens" of slime or within tubercles (see Figure 15) can make control very difficult. Treatments which have been recommended include various disinfection, acidization and cleaning practices. It is important when applying these various practices to ensure that: (1) the system has been flushed and cleaned as best as possible before starting the treatment program; (2) there is application of the highest recommended dosage and the longest contact time for the selected treatment program in order to maximize the potential effectiveness; (3) increasing the dissolved oxygen in the water can suppress SRB activity; and (4) there is consideration of the ongoing or routine application of disinfectants and/or penetrants to reduce the rate of recovery of the SRB from the "shock" effect of the practice (2).

The routine monitoring for the redox (Eh) potential may form a useful monitoring technique. Where the Eh in the water dramatically declines (i.e., becomes more reductive), this may be taken as an "early warning signal" that conditions are now changing and becoming potentially more supportive for another SRB outbreak. There are also a number of simple cultural systems that can be used to monitor for the presence of SRB organisms in water. It may be convenient to routinely monitor the water (e.g., monthly) for the aggressivity of these bacteria.

Other Sulfur Bacteria

There are some other groups of "sulfur bacteria" which may, on some occasions, cause problems in waters[29,66]. They may be summarized as the sulfur-oxidizing bacteria (Section 20), the colorless sulfur bacteria (Section 23), and the purple and green sulfur bacteria (Section 18). Of these groups, the sulfur-oxidizing bacteria are the most well documented because of their association with the recoveries of metals through the leaching of ores and the problems associated with acidic mine tailings. These bacteria require oxygen to grow and they convert various sulfides to acidic products such as sulfuric acid. The colorless sulfur bacteria do not usually produce acidic products but do convert hydrogen sulfide and other sulfides to sulfates. These bacteria are sometimes found growing in sulfur springs water wells and distributions systems where sulfides are present but where there is also available free D.O.

Oxygen is toxic to the purple or green sulfur bacteria. Like plants, these bacteria are able to photosynthesize but without the production of oxygen. Unlike plants, however, these bacteria reduce various sulfates to sulfides and elemental sulfur which becomes deposited in and/or around the cells. Common habitats for these bacteria include septic ponds where they occasionally dominate and turn the water red, and deeper lakes which have stratified (layered). These bacteria form distinctive plates of floating growth within the strata.

The various sulfur bacteria can be summarized therefore as belonging to the following major groups and genera:

Sulfate-Reducing Bacteria (SRB)
      Desulfovibrio / de-sul"fo-vib're-o
      Desulfotomaculum / de-sul"fo-to'mac"ool-um

Sulfur-Reducing Bacteria (SRB but uses sulfur)
   Desulforomonas / de-sul"f-ur-o'mo-nas
Sulfur-Oxidizing Bacteria (producing acidic products)
   Thiobacillus / thi'o-bah-sil'us
Colorless Sulfur Bacteria
   Beggiatoa / bej'je-ah-to'ah
   Thiothrix / thi'o-thriks
Purple and Green Sulfur Bacteria
   Chlorobium / klo-ro'be-um
   Chromatium / kro-ma'te-um

Archaebacteria (Section 25)

These are bacteria-like organisms that do not have the normal types of cell wall or genetic mechanisms found in the rest of the bacterial kingdom. Archaebacteria tend to be found in some of the more extreme and unusual habitats such as swamps, acid springs and salt lakes[12,62]. While there is a considerable diversity amongst this microbial group, there are in reality only three major groups known. These are the methanogenic archaebacteria (able to generate methane), extremely thermophilic archaebacteria (able to grow under conditions of high temperature), and extreme halophilic archaebacteria (able to grow in concentrated salt solutions).

Methanogenic archaebacteria produce methane (natural gas, biogas) anaerobically by converting simple compounds such as CO_2, H_2, formate, acetate to either methane or a mixture of methane and CO_2. These bacteria thrive in a wide variety of anaerobic environments rich in organic materials. Well-documented examples are the rumen, gastroenteric tracts, anaerobic sludge digesters and lake sediments. Some methanogens even parasitize other microbes such as protozoa. The common occurrence of methane in groundwaters would indicate that the methanogens may also be active in anaerobic niches within the aquifer, particularly where there is organic decomposition occurring. These microorganisms may well be the generators of the methane (CH_4) in landfill operations. Where these conditions are replicated in the laboratory setting, the gas can sometimes be observed to be biologically entrapped as separated bubbles with a biofilm forming the interface between the gas and aqueous phases. These bubbles then form into a dispersed foam spread through the porous medium. Such occurrences may cause a reduction in the hydraulic conductivity through the affected zone thus creating a temporary biological barrier. Where methane is being biogenerated, it can be expected that the methanotrophic (methane-utilizing) bacteria are likely to be particularly active at the aerobic (oxidative) interfaces. Common genera encountered among the methanogens includes *Methanobacterium, Methanothermus, Methanococcus, Methanomicrobium* and *Methanosarcina.*

Extreme halophilic archaebacteria are defined as those bacteria which require at least 8.8% NaCl for growth, and usually require 17 to 23% NaCl to achieve a maximal (optimal) growth. Higher salt concentrations up to saturation (32 to 36% NaCl) will gradually retard the growth of the extreme halophiles and some will still be able to grow albeit slowly at saturation. Two genera of aerobic bacteria (*Halobacterium* and *Halococcus*) are well known to be able to grow under these conditions where there is a sufficient organic content. These bacteria frequently produce bright red pigments which can color the water (seawater evaporation ponds often turn red where there are halobacterial activities).

These halobacteria are not the only microorganisms which can grow in salt-rich environments. Algae (*Dunaliella*) and phototrophic purple sulfur bacteria (*Ectothiorhodo-*

spira) can both thrive in surface saline water lakes with considerable primary production of organic material which then supports the halobacteria. In saline groundwater (> 8.8% NaCl) it can be expected that under aerobic (oxidative) conditions the halobacteria would be active in proportion to the available organic substrates. Under more alkaline conditions (pH between 9 and 11) where there is a low concentration of magnesium (Mg^{2+}), the dominant halobacteria may shift to two other genera (*Natronobacterium* and *Natronococcus*). While little research has yet been directed at the presence and activity levels of halophilic bacteria in extremely saline groundwaters, parallel studies on surface saline waters would indicate that microbial activity is likely to be present depending upon the availability of organic material.

Extreme thermophilic archaebacteria generally grow over a temperature range of 30C° with the minimum for growth being within the range from 55 to 85°C and maxima at between 87 and 110°C. Incredibly some of these bacteria can grow not only at these high temperatures but also under acid conditions (*Sulfolobus* and *Thermoplasma*). Most isolates have come from submarine and terrestrial volcanic sources but evidence would suggest that some thermophiles may be able to function in geothermally heated groundwater systems. For example, *Staphylothermus* has been found to be widely distributed in thermally heated seawater around "black smokers" (marine depth, 2500 meters) and it is thought that this genus may also be widely distributed near hot hydrothermal vents.

From the present knowledge of the activities of these archaebacteria, it is highly probable that many of the groundwater systems thought to be "hostile" to microbial activities due to high temperatures, salinity or acidic conditions may, in fact, be able to support the activities of microorganisms such as the archaebacteria.

Gram-Positive Cocci (Section 12)

There is a wide range of diverse types of gRAM positive coccal bacteria. Coccal indicates that the bacterium is spherical, ovoid or ellipsoid in shape. Genera which are composed of coccal bacteria usually bear the suffix -coccus. Aerobic cocci are dominated by the genus *Micrococcus* (mi"kro-kok'us) which occur in many natural habitats where oxygen is available. This range of habitats includes both surface- and groundwaters and micrococci are sometimes incorporated into the consortia of bacteria involved in biofouling events (e.g., plugging of a water well). Despite the frequency with which these bacteria are found, even on mammalian skin, only a few species can be pathogenic[10]. Another (strictly) aerobic coccus of interest is *Deinococcus* (dee"no-kok'is) which occurs widely in waters, ground meat products and feces. These coccal bacteria have an unusually high resistance to both desiccation and radiation effects. Where there is a saturated oxidative eutrophic regime being subjected to sublethal doses of gamma radiation, it may be expected that *Deinococcus* could become a major component in the consortial biofouling.

An alternative indicator bacterial group for hygiene risk due to fecal contamination is the streptococci referred to as the **enterococci** (group 3). These bacteria are normal residents of the gastroenteric tract of humans and most other animals and commonly appear in the feces. Sometimes the **enterococci** may be referred to as fecal streptococci including such species as *Streptococcus faecalis*. The ratio of **fecal streptococci** (FS) to **fecal coliforms** (FC) in water is sometimes used to determine the likelihood that the fecal material is of human origin. Where the water has been contaminated with human originated fecal material there is usually a preponderance of FC and the FC:FS ratio would favor the FC (e.g., 4:1 or 1:0.25). On the other hand, if the fecal contaminant had been of non-mammalian origin, the dominance would commonly shift to the FS group (e.g., FC:FS ratio of 1:4 or 0.25:1). *S. faecalis* is another opportunistic (nosocomial) pathogen which can cause urinary tract infections and endocarditis. Normally, FS do not persist for long in groundwater, but under brackish and saline groundwater conditions, the enterococci are likely to survive since they are able to tolerate and grow in 6.5% sodium chloride under suitable environmental conditions.

Endospore Forming Gram Positive Rods and Cocci
(Section 13)

Some bacterial cells form a very heat resistant body within cells where all essential components of the cell are concentrated. Once the body (spore) has matured, the cell material around it (hence the prefix endo- meaning inside of the body) is essentially dead and disperses. This leaves a small spore body which resists not only heat but also desiccation and other radical environmental shifts. These endospores are basically dormant viable entities. Some recent extrapolations suggest that the endospores of *Bacillus subtilis* may be capable of surviving between 4.5 and 45 million years in an interstellar molecular cloud while resisting the high vacuum, low temperatures and UV radiation present[62]. This may be an interesting example of the potential for survival that the endospore has under adverse conditions. Similar events may be expected to occur in groundwater systems where the dormant endospores may travel over considerable distances through millennia of time within aquifer systems.

Endospore forming bacteria are divided into a number of genera. Three of these are of particular interest in relation to groundwater environments. Of these, the most ubiquitous are the aerobic endosporogenous group called *Bacillus* (bah-sil'lus). Many species of *Bacillus* occur naturally in oxidative groundwaters and are frequently components in biofilms that cause plugging. The only species pathogenic to humans is *Bacillus anthracis*, the causant organism for anthrax. This species is highly infectious to animals and humans, and is passed along through direct contact. Its endospores can remain viable in soils for decades but there is little evidence of *B. anthracis* endospores surviving in or causing a threat from groundwater sources. While most species of *Bacillus* appear to be harmless to humans, there is a range of species that are pathogenic to insects (e.g., *B. thuringiensis*) and these often involve the generation of toxic agents lethal to the insects by these bacteria.

A second major genus in this group comprises the anaerobic endosporogenous bacteria: the genus *Clostridium* (klo-stri'de-um). This genus includes a diverse range of specialized degraders able to efficiently break down cellulose, chitin, proteins and other organic material under anaerobic conditions. Oxygen is lethal to most species in this genus and so their habitat is frequently restricted to organically rich saturated niches. The presence of *Clostridium* species in groundwater samples would indicate that the water was probably anoxic (contains no significant oxygen; < 20 ppb) and had tracked through a zone of relatively intense organic decomposition. Another source of organic material could be fecal contamination. The third major bacterial indicator species for hygiene hazard is *C*.

welchii due to its common occurrence as an inhabitant of animal gastroenteric tracts and fecal materials. A range of species of *Clostridum* produce powerful toxic agents which can affect man. These include *C. tetani* (tetanus) and *C. botulinum* (food poisoning, botulism). Neither of these species commonly use groundwater as a vehicle of infection.

Some sulfate reducing bacteria (SRB) possess endospores and hence are included in section 13 under the genus, *Desulfotomaculum* (de-sul"fo-to'mac"ool-um). Species of this SRB are frequently isolated as a component in consortia associated with corrosion.

Filamentous Actinomycetes (Ak"ti-no-mi'sez, Volume IV)

There is a large group of gRAM positive bacteria which form thread-like structures that the cells turn into filaments. These filaments may or may not branch. During successful growth, these filaments may form into a ramifying network called a **mycelium** (a common feature of growth among another microbial group, the Fungi). The nature of this growth is similar to the fungi in that the various filaments form an integrated network, usually by attaching to solid particles and forming networks between them. These bacteria commonly produce exospores (sporing bodies formed outside of the bacterial cells) which are not so resistant to harsh conditions as the endospores nor as capable of prolonged survival.

In groundwater systems, these filamentous actinomycetes are most commonly found at interfaces between saturated and unsaturated zones where oxygen is present together with adequate water. One focus site for such growths would be the capillary zone immediately above the saturated zone. Rates of growth would be controlled by the availability and type of organic materials present at the site either via the groundwater or such recharges as may be occurring. While some actinomycetes are anaerobic many are aerobic. The largest group of actinomycetes which are associated with groundwater belong to the genus *Streptomyces* (strep"to-mi'sez). Species of this genus form a very aggressive component within a microbial community. Many species produce various antibiotics such as the tetracyclines and aminoglycosides (e.g., streptomycin) which give these bacteria a competitive edge. Some species of *Streptomyces* produce earthy odors during their growth. In fact, the musty odor of soil is produced by *Streptomyces*. The compounds produced by these bacteria which cause this include the geosmin and isoborneol groups. Where an earthy musty odor is detected in a groundwater sample, an infestation by species of *Streptomyces* can be suspected.

Fungi

Some microorganisms have a more complex cell structure, larger size in filamentous (mycelial) form, and often exhibit distinct asexual and sexual reproductive cycles (among other distinctive features). These make up the Fungi. These microorganisms are mostly but not entirely aerobic and play a very major role in the degradation of complex organic material into simple organic compounds and inorganic molecules (mineralization). Frequently the fungi dominate biodegradation in unsaturated porous media and on organic surfaces under aerobic conditions. For example, in soils the fungi dominate the microbial biomass, which has been estimated to be 5 metric tons/hectare (2.2 tons/acre) in the top 20 cms[12]. Fungi are involved in a wide variety of plant diseases (5,000 species of fungi are pathogenic to specific plants of economic importance) and also the deterioration of manufactured goods which contain organics and are exposed to the air and moisture (e.g., fabrics, leathers, paper and wood products). In groundwater systems, the most likely site for fungal activity would be in an unsaturated zone where there are organic materials present. Examples of this would be a recharge zone from a surface sewage oxidation pond and in the capillary zone above a low density gasoline plume polluting a groundwater system. Working with fungi involves a different approach to scientific discipline, as compared to the other aspects of microbiology. The study of fungi is called **Mycology** (Greek *mykes*, mushroom, and *logos*, discourse). The different divisions and genera within the fungi are most easily differentiated by the form of cellular growth and reproduction that occurs. Very commonly, it is the manner by which the exospores (asexual reproduction) and zygopores (a form of sexual reproduction) are produced and the environments within which mycelia are created which is used in the classification of the fungi. Generally speaking the fungi can be differentiated by microscopic and cultural techniques while the bacteria are differentiated by staining, cultural and biochemical methods.

The Algae

Algae are some of the simpler members of the plant kingdom. They are all capable of photosynthesis using chlorophyll. Some algae are very large in bulk although they retain a simple form of tissue organization. These are the seaweeds (macro-algae) and are not found too often growing in wells! The simpler and smaller algae (micro-algae) are, however, sometimes recovered in water well samples. On occasions it is important to determine the origin and fouling potential relatable to these microorganisms.

In practise, the micro-algae can be relatively easily

divided into a number of major groups, in part by the color generated in the growth. These include: (1) the blue green micro-algae also known as the cyanobacteria; (2) the (grass) green micro-algae referred to as the chlorophytes; (3) the brown micro-algae called diatoms and desmids; and (4) the euglenoid micro-algae which are sometimes green or yellow and they may even be colorless as they shift between a "plant" form of existence and an "animal" form. The cell wall varies considerably between these forms of micro-algae which would mean that different control strategies would have to be used, depending upon the type of micro-alga recovered. Cyanobacteria are, as the name implies, bacterial and have a typical gRAM negative type of cell wall. Chlorophytes have a more typical cellulosic form of cell wall commonly found in higher plants. For the diatoms and desmids, the cell walls are often complex and highly structured (ornate) with a high silicate content. Euglenoids possess a cell wall more typical of protozoa (i.e., animals) and are more complex and dominated by proteins.

All of the micro-algae are capable of photosynthesis even under the very low levels of light that may be found in some wide bore shallow wells or in the soil. By way of a comparison, in soils the micro-algae appear to be able to grow phototrophically down to depths of 50 mm or more. The cyanobacteria tend to dominate in a horizon from 5 to 25 mm with the other micro-algae growing at greater depths. It is becoming apparent that many of the micro-algae are also capable of growing heterotrophically, that is, competing with the mass of bacteria, fungi and other organisms for organic nutrients. This is particularly evident for the euglenoids and the chlorophytes. The occurrence of these micro-algae in well waters which have not been subjected to any level of light therefore becomes less surprising. Another mechanism by which micro-algae may enter a water well (particularly a shallow wide bore or lateral gallery well) is with the recharge water coming from soil or surface water reservoirs. From soils there is thought to be a continual "bleeding" not just of micro-algae but of the microflora at large. These microorganisms (many possibly as UMBs) move downwards with the recharge waters to enter and move into the groundwater systems. Little is known of the distances over which these movements can be achieved.

The presence of micro-algae in a well water sample should raise a sequential series of questions:

1. Is there light penetrating into the water column of the well? Yes/No. If No goto 3.
2. The probability is that the algae are growing in the well. If there is a low total organic carbon but relatively significant levels of inorganic nitrogen (e.g., nitrate, ammonium in low ppm range) and phosphate (in the high ppb range), then algal growth can be expected.
3. The micro-algae are either competing heterotrophically for organic carbon with other microorganisms or they have "contaminated" the well water with surface recharge. Is the total organic carbon greater than 2.0 ppm? Yes/No. If No goto 5.
4. There may have been either a heavy algal population in the recharge water or the algae are competing heterotrophically for the organic carbon.
5. There is a likelihood that the algae present in the well have arrived via recharge water originating in soil or surface waters which supported algal growth.

6

The Management of Microbially Induced Fouling

DEFINITION

Microbially induced fouling (**MIF**) within groundwater systems is directly induced by microorganisms operating either within the porous media of the aquifer or at the boundaries. Such microbially induced biofouling may yield a mixture of benefits and costs to the users.

Benefits of **MIF** generally focus on the improvement in water quality which may occur at the biologically active interfaces (biozones) as a result of bioaccumulation and biodegradation. Such improvements may only occur for a limited period of time and be terminated by the inhibition, terminal maturation or saturation of the biozones. Other benefits may also accrue through the bioimpedence of flow from zones within the aquifer where the groundwater may be less acceptable to the user (e.g., a contaminant plume). Here, the biological activity may take the form of semipermeable or impermeable barrier (biobarrier) which performs two major functions. These are: (1) degradation and/or the bioaccumulation of the targeted pollutant, and (2) a partial or total impedance of transmissivity (flow).

Costs to the user of **MIF** are multiple. These range from corrosion of the installation (microbially induced corrosion, **MIC**), through to steady reductions in production (due to bioimpedence) until a plugging of the water well occurs, to deteriorations in water quality as the bioaccumulated chemicals become released in the maturation cycle of the biofouling[21,22,50].

Management of an **MIF**, when it is recognized to be occurring in water supply, recovery, extraction, injection or monitoring wells, should be performed in such a way as to maximize the benefits and minimize the cost. Since the **MIF** is essentially a natural event (i.e., involving microbial components which occur commonly within the crustal and surface biospheres), each particular event may be viewed as being unique in its nature and therefore more difficult to manage. In spite of this restraint, there are two major approaches which can form a part of a management strategy. These include suppressive and supportive approaches. A suppressive approach would have the primary objective of eliminating the deteriorations (e.g., corrosivity, plugging or water quality declines) within the system under the users control. A supportive approach less commonly practiced would have the primary objective of manipulating the **MIF** to achieve a maximum of benefits (e.g., improved water quality and productivity). Both the suppressive and supportive approaches require an understanding of the various processes and factors which affect the activities of microorganisms if management is to be achieved.

SUPPRESSIVE STRATEGIES

These are designed to generate a temporary or radical reduction in the microbial activities and presences within a system that are believed to be causing a deterioration. Unless the infested zone can be totally isolated from its surrounding environment, there is a strong probability of reinfestation and a recurrence of the deteriorations. These strategies are therefore suppressive in nature and may not normally be considered to be permanent solutions to the concerns. A **temporary suppressive** event normally will use chemical and/or physical strategies to traumatize (shock) the biota and temporarily disrupt the deterioration process. A **radical suppressive** event would involve a prolonged or more intensive use of chemical and/or physical strategies to destroy much of the biota and eliminate the deterioration processes for an extended period of time. The length of this extension period would be inversely relatable primarily, but not exclusively, to the rate of recolonization of the destroyed habitats.

Temporary suppressive strategies

In these strategies, the objective is normally to control the deterioration event just sufficiently to allow the product water to remain acceptable and/or the installation to function efficiently. Generally these strategies involve either a single or a paired chemical and/or physical intrusion with a maximum of convenience (e.g., the discharge of the suppressive chemical directly down the water column within the well)[26]. The two major strategies involve the use of disinfectants (chemical therapy) or the application of chemicals to cause a significant shift in the pH of the water (e.g., acidization)[73].

Chemical therapies primarily rely upon a disinfection being achieved[84]. The objective here is to destroy the "hygiene-risk" and "harmful" microorganisms whether these organisms are in a planktonic, sessile particulate or sessile attached phase. In the application of the disinfectant, consideration has to be given to the rate at which the applied chemicals can penetrate the protecting ECPS layers encompassing the sessile microorganisms and then "attack and destroy" the incumbent organisms[51].

Three major therapies have been developed to increase the effectiveness of the disinfectant chemicals. These are: (1) increasing concentration and contact time; (2) application of surfactant chemicals which will disrupt the ECPS coatings and cause a more rapid intrusion into the targeted organisms; and (3) coupling of two treatments together in such a way that the resultant effect is greater than could be achieved by applying each component separately (i.e., a synergistic effect).

One major problem which has arisen in the application of these treatments is the difficulty with which the impact of the treatment can be gauged. Microbiological analysis of the postdiluvial water after such treatments may not reflect the level of impact that the treatment may have had on all of the targeted groups of organisms. For example, where disinfectant X was applied, there is a high probability that the most vulnerable group of microorganisms would be those in the planktonic state. It would therefore be probable that these organisms would be eliminated by the treatment. However for the sessile organisms there would have been some level of protection afforded by the presence of coatings of ECPS and bioaccumulated minerals. For the sessile particulates which are suspended in the water there is a high probability of many of the incumbent organisms being killed while many of the rest would be in a state of post-treatment shock (trauma). Test methods which rely upon the growth and/or metabolic activity occurring within a fixed time frame may indicate a zero activity because the surviving organisms are in a state of shock. A prolongation of the test period may reveal the survivors once the period of trauma has ended.

Of the three groups of microorganisms, it would be the sessile attached organisms which would be the most difficult to monitor. These organisms occupy niches within an attached biofilm and so present a smaller surface area to the applied chemical and/or physical treatment. One common reaction of these communities when subjected to severe chemical and/or physical stress is to contract in size. This may be caused by the expressing of "bound" water held within the ECPS to the "free" water state. A consequence of this is that the density of ECPS in the remaining volume of biofilm rises and impedes the diffusion of chemicals through the polymeric matrices to the incumbent cells.

Another reaction relates to the impact of shifting pH in the surrounding water. Where the pH of the water is suddenly artificially lowered (e.g., such as at the initiation of acidization), the biofilms do have some ability to "buffer" the impact and even return the pH over a period of contact time to the original levels. Such a buffering capacity within the biofilm would act to counteract any impacts that the pH shift would have on the incumbent microflora within the biofilms comprising the fouling.

Monitoring the impact of a given treatment upon the incumbent microorganisms comprising the fouling would therefore tend to be flawed because of the following factors: (1) elimination of much or all of the viable planktonic forms; (2) traumatization of the sessile particulate viable entities so that false low values may be obtained; and (3) failure to register the impact of the treatment on the sessile attached microflora. In the latter case, where shrinkage of the biofilms is a response to the treatment event, faster flows may be recorded and zero or very low numbers of suspended viable entities. Both of these events would indicate that the treatment had been perhaps more successful than had, in reality, been achieved. This will usually be seen in the lack of longevity in treatment effectiveness. Clearly it is only through prolonged and careful monitoring that the true impact of a given chemical and/or physical treatment can be ascertained.

Disinfection, this is a descriptive term for the application of an antimicrobial agent to inanimate objects in such a way as to destroy "harmful" microorganisms. The term, harmful, can relate to humans (as potential infection risks) or to the functionality of the inanimate object (i.e., the production efficiency of the water well). Classes of disinfectant chemicals include phenols and phenolics, alcohols, halogens, ozone, surfactants, alkylating agents and heavy metals. Physical techniques are more commonly associated with sterilization (absolute elimination of life forms) rather than disinfection (differential elimination of life forms) but can be used for disinfectant functions. These include the use of heat, filtration, ultrasonics and radiation (electromagnetic, ionizing and ultraviolet).

The "ideal" qualities for a disinfectant[10,62] has generally been considered to involve the following qualities:

1. Able to destroy all forms of microorganisms, including spores, within a reasonable period of time.

2. Would not react or be interfered with by natural materials present at the site of disinfection. In reality, disinfectants tend to interact with the reactive groups of organic matter such as occur in proteins and biofilms.

3. Would be nonallergenic, nontoxic nor irritating to tissues.

4. Should be noncorrosive, nondiscoloring and not degrade at a rate which would influence effectiveness. Produce no unsightly products or toxic residues.

5. Effectively contact the targeted organisms because of wettability and penetrability.

6. Would be soluble (preferably) in water or common organic solvents.

7. Chemically stable and should, where in a liquid form, not be subject to evaporation which could create a too-concentrated solution which may become damaging to tissues and objects.

8. The effective range would not be bypassed as a result of a reasonable diluting factor.

9. There would be no disagreeable odor, no residual stains,

nor would it be inordinately expensive.

Clearly it would be very difficult for a single disinfectant to meet all of these conditions. Disinfectants must therefore be chosen on the basis of the job that is at hand and the intrinsic properties of the chemical. Each individual form or group of disinfectants perform the disinfectant function in different ways and, in consequence, are used in specific ways.

Phenol and Phenolics. Phenol is commonly considered the classic medical disinfectant. It was introduced by Joseph Lister as a means of reducing surgical infections. The form used was carbolic acid, which was applied in aerosol sprays or in saturated surgical dressings. The success of Lister's technique focused attention on phenol and phenolic compounds as potential disinfectants. This group was found to have many favorable qualities. Compounds would remain active even in the presence of organic material, are stable and persistent after application and, when dried, will become active again when rehydrated. Unfortunately, the phenol and the phenolics were found to have a number of undesirable side effects which have restricted their use to sanitization functions. For example, hexachlorophene (HCP) was widely used in the control of hospital nursery infections (e.g., staphylococcal). Experimental studies however showed a greater risk of brain damage where newborn infants were exposed to three or more daily baths in 3% HCP. These and parallel findings have led to the restricted use of this group of disinfectants. This class of chemical disinfectants would be considered undesirable in well cleaning since these are classed as groundwater contaminants. Corrosivity is another concern for this group of disinfectants as well as the tendency for these chemicals to cause skin irritation.

Alcohols. Ethyl and isopropyl alcohol are two extensively used alcohols but function at concentrations in the 60 to 95% range. They are effective against broad groups of bacteria, fungi and viruses but are not sporocidal. Ethyl alcohol has been found to be effective against microbial activity on environmental surfaces where high concentrations have been used even where these surfaces are biofouled. Generally, the effectiveness of the treatment declines rapidly where the concentration of the ethyl alcohol has been reduced to below 45%.

Halogens (chlorine, iodine and bromine). Halogens are usually employed in inorganic forms. Their mode of effective action is due to oxidation functions and the direct halogenation of proteins causing radical cellular dysfunctions. While all halogens do have some use as substitutes in other forms of disinfectants, the commonly used halogens are chlorine and iodine. Bromine compounds are used extensively for the "shock" treatment of cooling tower slimes.

Elemental chlorine and inorganic chlorine compounds are employed widely in the water industry for sanitization, purification and disinfection functions. Of these forms, the most commonly employed are: chlorine dioxide (ClO_2), chlorine (Cl_2) and hypochlorite (OCl^-). While the antimicrobial function does operate in the ppm range, organic matter does interfere with this activity. It is therefore generally accepted that while there remains a "residual chlorine" presence there will continue to be an antimicrobial (disinfecting) function. Sodium hypochlorite ($NaOCl$) is one of the commonest disinfectants. Normal concentrations for a household strength bleach is 5.25%. Surface disinfection uses dilutions of this bleach ranging from 1:10 to 1:100 and has to be freshly prepared on a daily basis. The precise ratio would be directly influenced by the amount of organic material present on the surface. These concentrations can be corrosive to metals, particularly aluminium. With groundwater supplies, a disinfection is commonly achieved at 1-3 ppm dosage rates while for water well disinfection, dosages are commonly elevated to the 200 to 1,000 ppm levels. Chloramines and quaternary ammonium compounds (QAC) are also used on occasions. These may also contain available chlorine radicals to perform a disinfectant function.

Elemental iodine and iodine compounds are currently used as a general disinfectant in the medical industry (e.g., thermometers, surgical appliances and for wound or skin antisepsis) and for the purification of water. Iodine in the elemental form is relatively insoluble in water (saturation in cold water, 290 ppm). Where iodine is used in the purification of water, it is common for a bypass line to be passed through an iodine matrix. The iodine-saturated water is then mixed with the mainline water in a ratio to allow the desirable level of iodine to be carried. In medical practices, iodine is used in a hydroalcoholic solution as a tincture which is effective in destroying skin bacteria. Iodine at high concentrations can cause tissue necrosis but the effectiveness of iodine as a disinfectant has generated a considerable research effort. The major group of compounds generated by this activity are the iodophors. These are complexes of iodine and surface active organic carrier molecules. While the activity levels are lower than could be achieved by iodine, these iodophors are nonstaining, nonallergenic, water-soluble, and relatively nonirritating.

Ozone. Ozone (O_3) is a gaseous entity generated by the exposure of oxygen (O_2) to strong electrical discharges. Ozone acts as a very strong oxidizing agent and is used to purify drinking water. It is, however, an unstable molecule which decomposes back to oxygen. At room temperatures, the half-life (50% decomposition) of ozone is fourteen hours in air. This instability inherent in ozone makes it impractical to use as a compressed gas and so it is generated at site. Commercial ozone is generated by accelerating electrons between two highly charged electrodes. An interaction takes place between the electron and oxygen molecules causing dissociation to two oxygen atoms. These atoms interact to form ozone. It is the very high oxidation potential of ozone (-2.08V) that makes ozone such a powerful disinfectant. Chlorine, by comparison, has an oxidation potential of - 1.34V and fluorine (most powerful oxidation potential known) is -2.87V.

Ozone is a colorless gas. Commercial production of ozone from air usually yields 2% by weight. Before a disinfection of microorganisms in water can be accomplished, the ozone has to be transferred from the charged air to the water phase. This may be achieved using some form of ozone contacting device in which the ozone residual after passing through the contactor could reach 0.9 to 1.0 ppm. The instability of the ozone molecule and the aggressive oxidative nature of the chemical causes rapid declines in the ozone concentration particularly where there is a considerable presence of organic material (e.g. where biofouling is occurring). This leads to very rapid reductions in the residual ozone concentrations and limited ability of the ozone to penetrate effectively into biofilms. The greatest effectiveness of ozone is therefore against suspended planktonic microorganisms. A concern in the used ozone is that the oxidation processes may yield additional amounts of degradable assimilable organics to water. This could lead to an increase in microbial activity downstream from the treatment site.

Surfactants. These compounds are surface-active agents which have the property of concentrating at interfaces. The effects include the lowering of surface tension, solubilizing and wetting of the impacted compounds. Soaps and detergents both have surfactant properties. Soaps are sodium or potassium salts of long-chain fatty acids and generally act in a "cleansing" manner removing microorganisms from surfaces such as skin but often with little antimicrobial activities. Bactericidal agents are sometimes added to soaps (50% of cosmetic types) in order to generate some direct antimicrobial activity. Detergents are generally considered to be superior to soaps as cleansing agents. Unlike soaps, detergents are synthetically produced surfactant cleansing agents. Detergents are regarded as superior to soaps because they do not form precipitates or deposits with water minerals whereas soaps do. There are three major types of surfactant: anionic, cationic and nonionic. These are separated by the polarity and activity of their polar groups.

Anionic detergents. They are not disinfectants but, like soaps, do remove microbial contaminants with grime and dirt. Most of the popular disinfectants, antiseptics and sanitizers in this group come from the **cationic detergents.** These (cationic) have positively charged components of the lipophilic alkyl chain portion of the molecules which are attracted to the negatively charged membranes in the microbial cell wall structures. The detergent causes a lethal disruption of these membranes. Even enveloped viruses which also possess these types of membranous structures are inactivated by cationic surfactants. Three major cationic detergents are popularly used for their antimicrobial properties: *benzalkonium chloride, benzethonium chloride,* and *cetylpyridinum chloride.* Cationic detergents are bland, very stable, nontoxic, inexpensive and remain active at high dilutions. However, these advantages are counterbalanced by the lack of sporicidal activity, resistance of some pseudomonad bacteria (at high dilutions) and ready neutralization by organic material, soaps and anionic detergents.

Alkylating Agents. These exert an antimicrobial effect due to the ability of the agents to substitute alkyl groups for hydrogen in the reactive components of enzymes, nucleic acids and proteins. Common components which contain the labile reactive hydrogen are the carboxyl (-COOH), sulfhydryl (-SH), amino (-NH$_2$) and hydroxyl (-OH) groups. There are three major agents; formaldehyde, glutaraldehyde and ethylene oxide (ETO).

Formaldehyde can be used in the gaseous state as a fumigant or in solution (formalin). It is not widely used as a disinfectant or sterilant because it is a tissue irritant. Major applications are as a major component in embalming fluids and as an inactivator in viruses.

Glutaraldehyde is categorized as a *dialdehyde* and has recently been introduced as a disinfectant. It has a wide spectrum of activity and is effective at the 2% solution strength against bacteria and viruses (10 to 90 minutes) and spores (10 hours). Unfortunately, this compound is very irritating to the skin, mucous membranes and the eyes which restricts the application.

ETO is a gaseous cyclic ether which is able to sterilize in the absence of high levels of heat or moisture. A wide range of heat sensitive equipment (e.g. disposable laboratory items, textiles) can be sterilized with ETO. It is toxic if inhaled and will cause blisters on contact. Generally ETO is slow to function as a sterilant taking 6 to 12 hours at room temperature. The hazardous nature of ETO has restricted its use to specific controlled environments. For example, many hospitals now have ETO autoclaves as an additional method for sterilization. The ETO method does not have application under normal circumstances to groundwater systems.

Heavy metals. Most heavy metal disinfectants are based on preparations containing mercury or silver in either inorganic or organic forms. Most have now been superseded by other forms of disinfectants which are more effective and less of an environmental hazard. Silver nitrate (1%) is still used in limited medical applications (e.g., burn treatment, reduced risk of conjunctivitis in the newborn). The use of heavy metals for disinfection does not have application under normal circumstances to groundwater systems but may be collected (MIA) within biological interfaces (see Figures 3 and 9). When cleaning out a biofouled well after treatment, the slime may contain high concentrations of metals (through MIA) and need to be treated as a potentially hazardous discharge.

Summary. All of the disinfectants discussed are influenced in their effectiveness by the environment within which these chemicals are placed. Groundwater environments offer particularly challenging conditions since there would be a number of distinct environments within which the chemical could be placed. For example, a liquid disinfectant injected down through the water column of the well and then surged could impact on three major types of environment. These are: (1) water column with a low surface area to volume ratio and a high particulate population; (2) biofouled porous media zones where the surface area to volume ratio would be high and the water-surface interfaces would be principally composed of biofilm structures; and (3) nonfouled porous media which would also have a high surface area to volume ratio but where the surfaces would be "relatively" pristine with either thin or no biofilm present. The ideal disinfectant to remove the infestation from such sites would have a range of characteristics including: highly diffusible; able to penetrate rapidly through or destroy biofilms; effective against a wide range of microorganisms at very low concentrations; not be neutralized by inanimate inorganic or organic chemicals; be stable; slowly biodegradable; not generate any toxic or nuisance byproducts; be odorless, colorless, nontoxic, nonallergenic and nonreactive to humans, plants and animals. In reality, such a disinfectant forms an "ideal".

Of all of the disinfectants, it is the chlorine group that is the most widely used to control biofouling within groundwater systems. The principal advantage of this group is the effectiveness at low concentrations (i.e., low ppm range) to control suspended planktonic and (to a lesser extent) the suspended sessile particulate microorganisms. Chlorine disinfectants are particularly effective in controlling the coliform bacteria and are therefore widely accepted as a primary means of ensuring a potable water with an acceptable hygiene risk. A further major advantage is the ability of the user to determine whether there is any ongoing disinfectant potential. This is estimated by the amount of residual chlorine (i.e., still active chlorine) remaining in the water. There are, however, two major concerns in the application of chlorine based disinfectants to biofouled groundwater systems. These are: (1) the chlorine has little ability to penetrate a biofilm structure so that the action of the disinfectant against sessile attached microorganisms is slow and frequently incomplete; and (2) the oxidization potential of the chlorine is not target specific (i.e., the viable microbial cells) and much of the disinfecting properties may be neutralized by chance oxidations of dead organic materials. A further concern is that the chlorine, in reacting with organic compounds can generate *trihalomethanes (THM)* which may then enter the product waters. These THM compounds pose a health risk to the consumer when present in significant concentrations.

In applying chlorine to a biofouled porous medium, some of these disadvantages can be reduced by raising the concentration of the chlorine applied (e.g., *sodium hypochlorite* from 100 to 5,000 ppm) and extending the contact time (e.g., from 1 hour to 2 days). It should be remembered that the contact time may be cut short by the loss of a residual chlorine value in the water. Once this is gone then it may be assumed that the disinfecting action has also been curtailed. Furthermore this would indicate that there was possibly a considerable amount of organic material or inorganic minerals (possibly from a biofouling event) which was neutralizing the chlorine.

There are a number of strategies for improving the effectiveness of chlorine disinfectants. These include the coupled use of a surfactant (to disrupt the biofilm and allow faster penetration of the chlorine), or acidization where the hydrolysis of the biofilm structures may be sufficient to cause the system to collapse and allow disinfection to proceed efficiently.

Ozone is becoming a very widely applied disinfectant offering the advantages of no significant byproducts. Its instability in treated waters means however that there is little residual ozone left in the treated waters. This would allow any post-treatment downstream microbial infestation to flourish unabated by the treatment. The application of ozone treatment to severely biofouled porous media would have the same scale of difficulty as for chemical disinfectants. The large surface areas of biofouled material would provide a major barrier to ozone reaching and then destroying the incumbent microorganisms.

Radical Suppressive Strategies

There have been shown to be a number of distinct barriers to the successful disinfection and restoration of a biofouled groundwater system. Most significant of these "barriers" is the organic components of biofilms (slimes) which act both as neutralizers and retardants to the chemical disinfectants. In radical suppressive strategies, a major objective is to destroy these organic barriers by either physical or chemical means and remove the material as far as possible from the site of treatment (by techniques such as surging or pumping to waste). Physical strategies are being increasingly considered as potentially major components in these types of restoration. These include the use of temperature shifts, pH manipulation, ionizing radiation and controlling the REDOX potential.

Temperature shifts. Microorganisms growing within a groundwater environment are usually exposed to a temperature regime which is very stable. In shallow wells (i.e., <50 meters) which are being subjected to local surface water recharges, it can be expected that the maximum vacillation in temperature may be 10°C either side of ambient. However, where the well receives all of the water from a groundwater source, there may be expected to be a much narrower shift in temperature even through the seasons or from pump motor heating (i.e., <5°C). These microorganisms are therefore not subjected to the same extremes of diurnal or seasonal temperature changes that microorganisms in a surface water environment would experience. It may be expected that these organisms experience a temperature regime which exhibits relatively little variation when compared to surface water environments. Such relatively stable temperatures would indicate that the microorganisms growing within these niches would not be exposed to radical or even moderate diurnal (daily) or seasonal temperature shifts. Radical suppressive strategies can involve the shifting of the temperature outside of the normal ambient range to stress the biofouling microorganisms.

One radical suppressive strategy involves the application of heat (usually downhole) to elevate the temperature in the fouled zones to above that tolerable to the incumbent organisms. Louis Pasteur in the control of spoilage in beer, applied temperature trauma conditions by the holding the beer minimally at 60°C for 30 minutes. This technique (pasteurization) was later successfully applied to other food and beverage products (particularly milk). High temperature and short time (HTST) and then ultra high temperature (UHT) pasteurization processes followed, producing even better control of spoilage. Biofouling is a parallel event in that the product (e.g., potable water) is being "spoiled" by the activities of the various microorganisms in the delivery system (e.g., causal through to postdiluvial water). The principles of pasteurization can be applied to a water well through the application of heat to raise the temperature within the fouled zone by at least 40°C for a period of at least 30 minutes.

Experimental well restoration studies were conducted through the 1970s and 1980s using a variety of techniques (Figure 23) to achieve pasteurization. These have included: (1) steam injection into the water column; (2) recycling water from the well through a heat exchanger and returning the heated water to the well; (3) *in situ* heating of the water column (usually at screen level) with electric immersion heaters; and (4) direct microwave radiation heating using a downhole antenna. All of these techniques encountered a major obstacle which was the rate at which the thermal energy could move through the biofouled zone. Where severe occlusions were present, the heated water would tend to diffuse along the channels of least resistance (e.g., most open structures and/or rise on thermal convection currents. Longer time frames than that envisaged for a liquid product (e.g., milk) and the maximal tolerable temperatures to be applied have therefore to be considered.

The materials of construction for the water well may provide a barrier to the uppermost temperature that can be applied to a water well. Many plastic casings, for example, have not been designed to withstand temperatures of greater than 60 to 70°C. It would be foolish therefore to attempt pasteurization on a water well known to have components which could not withstand, for example, 65°C. Clearly, the

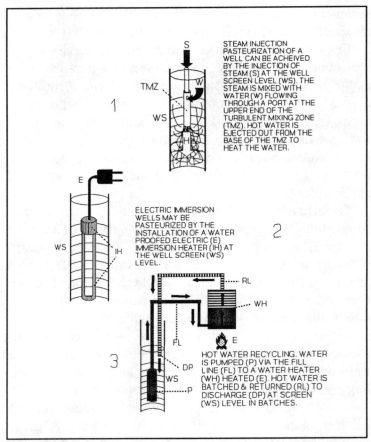

Figure 23. *Methods of pasteurization of a biofouled water well using: (1) steam injection (TMZ, patented BCHT process), electrical immersion heaters (2) or hot water recycling (3).*

higher the temperature that the structures of the water well can withstand then the higher the temperature that can be applied. Additionally, the longer the heat is applied to the well the further from the well the thermal effects will be felt until stabilization occurs (i.e., the thermal input matches the thermal losses within the treated zone and the temperature becomes stable). Once a thermal gradient has been established around a well, there will be a range of effects. Close to the well, the temperature may be sufficiently high to kill all of the vegetative cells and only some of the more resistant (endo)spores may survive. The slime layers would be severely disrupted and all of the incumbent cells impaired or killed. Beyond this zone of lethality, the temperature would be sufficiently high to kill the majority of the vegetative cells and severely disrupt the slime layer. Some survival may occur in deep seated niches where the heat fails to penetrate efficiently (e.g., at the interface between two large gravel particles). Further out from the zone of lethality there would be a gradually reducing temperature effect so that the percentile of surviving organisms would gradually increase due to higher temperatues but not in the lethal range while the slime layers would still be severely disrupted. The temperature range most likely to cause this partial kill and extensive traumatization of the incumbent microorganisms would be between 35°C and 50°C. From ambient groundwater temperatures up to 35°C, there would tend to be a differential shift in the microbial consortium with some being able to function more efficiently while others would be inhibited.

It can be projected that there are therefore four zones of impact that the thermal gradient may have on the incumbent microorganisms within the various biozones. In all four events there would be severe stress on the structures comprising the biofouling slimes. These stresses may create a number of product effects: (1) the structures break up into dispersed suspended biocolloids which shear away from the surfaces; (2) the structures partially shear with the lower strata of the biofilms compressing down; or (3) the structures "coagulate" to form a heat resistant gel of approximately the same volume which remains able to impede flow. In the events where the product effect is (1) or (2), there is a recovery in flow since the bioimpedence factors (i.e., the plugging slimes) have been drastically reduced in volume. In event (1) much of the organic material is turned to a suspended (from an attached) phase and may be pumped out of the well thus denying the residual microflora those (removed organic) nutrients for future growth. Where event (2) occurs, some of the organic materials will remain attached (albeit in a stressed and compressed form) which can form a nutrient base for future biofouling. When event (3) predominates, relatively little recovery in flow may be observed nor will there be a massive amount of biocolloids released in the post-pasteurization pumping. Even if all of the incumbent microorganisms were to have been killed in an event (3) the large amounts of available dead organic material contained in the thermally gelled slimes would form a major nutrient source for future microbial colonizers.

Each of these product effects may be expected to cause a different recovery/secondary fouling event (Figure 24). Secondary fouling would relate to the activities of colonizing microorganisms reoccupying the niches where the incumbent microorganisms had been destroyed by pasteurization.

Product event 1 would exhibit very heavy discharges of biocolloids in the post-treatment phase. Initially, there would be very few viable cells present in the discharge water but as water arrives from further out along the treatment gradient there may be large "surges" in the number of viable cells which have survived on the fringes of the thermal zone of lethality. These cells would, for the most part, be incumbent in the biocolloids formed as the slime layers shattered, sloughed and went into a suspended particulate state. With continued pumping, there would normally be a gradual decrease in the viable count until very low (background) numbers were recorded. Flows would return to close to original specified flows and prolonged recovery may be expected (subject to the nutrient loading inherent in the groundwater).

Product event 2 would exhibit heavy discharges of biocolloids for a relatively short period of time followed by periodic often radical variations in viable counts or dd as the residual compressed slime continues to slough erratically. Flows may initially recover close to original specified flows but these could be of short duration. This would be due to recolonization of the compressed slime layers which would be relatively rich in nutrients. Recolonization may occur so rapidly that the flows through the biofouled zone may become curtailed so that the water well exhibits all of the symptoms of being plugged again.

Product event 3 would exhibit much lower levels of discharge once pumping was started again and relatively little recovery in flow. While the volume of thermally gelled slime may be little different to that of the pretreatment living slime, it is a "gold mine" of nutrients for any microbial colonizers. In consequence of this usage it can be expected that the gelled status of the traumatized slime will rapidly revert to a very active biofouling and severe symptoms will again rapidly be displayed.

A successful pasteurization would be one that ended in product event 1 but even here there is a high probability that a recurrence of the biofouling will happen over time. Clearly where the treatment leads to product event 2 or 3 any gains in terms of water flow and quality may be short-lived. Where these events are observed clearly what is happening is that there is a rapid recolonization of the available spaces using the residual nutrients. It is not practical to physically remove these "catalysts" for future growth from the groundwater systems but there is one technique which has traditionally been used in the food industry to control persistent infestations. This technique is known as **tyndallization** and involves the application of three separate heat treatments spaced sufficiently far apart that any incumbent surviving microorganisms would enter into an

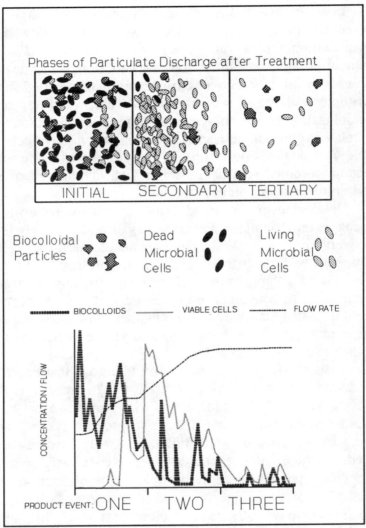

Figure 24. *Diagrammatic presentation of discharges from a traumatized biozone leading to the releases of particles (upper) and changes in flow rate and releases of particles (lower).*

active growth phase (and, incidentally, become more vulnerable to the next applied thermal effect). The spacing in the food industry was set at one or two days between treatments but in the groundwater environment (which has lower nutrient loadings than food) a longer time (e.g., seven days) should be allowed to elapse between treatments. This would allow enough time for the compressed or gelled slimes to become colonized and regrowths to occur. During these early growth phases, the biofilm structures tend to expand, become less dense and, consequently, more susceptible to becoming a product event 1 when heat is applied. Repeating this cyclic growth-treatment phase twice would cause much of the incumbent residual slimes to become looser and more completely removed in the pumping following the treatment phase.

Both pasteurization and tyndallization employ a combination of applied heat and time to maximize the stress to initiate sloughing of the slimes and the death of the incumbent cells. The precise temperature-time combinations are variable, subject to the size of the water well, the biofouled zone volume that will require treatment, and the ability of the well structures to survive the applied heat and the availability of a suitable heat source. Unless the heat source is extremely powerful and can bring the water column up to boiling point (100°C), there will be a temperature rise until equilibration is established. The structures of the well have to be both resistant to the applied heat and to the considerable turbulences that may accompany the heating if higher temperatures are to be reached and held. Field experience has found that equilibration (heat loss balances heat input) will usually occur within twenty-four hours of the start of continuous heat application. Upper limit temperatures reached may range from 45 to 100°C, depending upon the relative size of the well and the output from the heat source. Significant traumatization of the slimes and releases (sloughing) of particulate matter appears to commonly occur when the temperature rises above 50 to 55°C. Ideally, the whole of the zone of recognized biofouling (i.e., the distance from the midpoint of the well to the outer edge projected for the furthermost biozone) should be treated to a minimum of 60°C for 45 minutes. Biozones closer to the well would receive higher temperature and longer exposure times by the nature of the thermal gradient that would be established.

The selection of the type and size of the heat source would be dependent upon the size of the water well installation. A small water well suitable for an individual household (10 to 16 cms diameter and 30 to 40 meters in depth) may be pasteurized using either steam injection, hot water recycling or a screen level electrical immersion heater. Steam injection could easily be performed using a 300,000 btu/hr output small steam generator and the steam may be injected into the water column at the screen level. Temperature elevations could be very fast (>5°C/min) and the steam input may have to be throttled back to prevent boiling. Hot water recycling uses a hot water heater (e.g., 250 liters) or a swimming pool heater (modified to low flow and higher discharge temperatures). Water from the well is pumped into the heater, heated to a higher discharge temperature and returned to the well water column. The water heater uses a pulsed sequence in which water is pumped into the heater periodically and, by displacement, forces the hot water already in the heater to discharge down the well. This process is a slow heating cycle dependent upon the rate at which the water warms in the heater. Each treatment pulse may occur every 30 to 40 minutes when the hot water in the heater has reached 80 to 85°C. Displacement discharge then occurs in which the hot water now enters the water column. Temperature rises are very slow with the initial elevations reaching between 5 and 15°C and equilibration occurring after four to six hours (see Figure 25).

Waterproofed electric immersion heaters are convenient to install within the screen area but have a relatively low

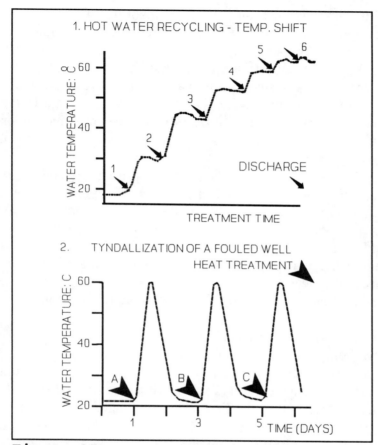

Figure 25. *Temperature profile (°C) of a water well being treated by hot water recycling (1, upper) or by tyndallization (2, lower). Upper arrows show times of hot water discharge into well.*

energy output compared to the amount of heat required where a major biofouling event is occurring. These systems therefore are more likely to prove to be satisfactory where the biozones are close to the well with a maximum diameter of fouling of two meters. Larger zones of fouling are not likely to be penetrated by an adequate thermal effect to destroy the slimes. Normally the limiting factor to the size of the immersion heater is the availability of power. For example, a 40 amp line would be required to operate an 8 kw heater (output: 27,312 btu/hr) while a 100 amp line would comfortably support a 16 kw heater (output: 54,624 btu/hr). This system lends itself well to the control of a localized biofouling event that would require tyndallization since the heat treatment process could be performed overnight with the discharge of the treated waters occurring in the early morning.

The straight use of heat as the primary rehabilitation mechanism may be flawed in cases where a product event 3 is a likely occurrence. Here, the thermally gelled slimes may prove difficult to destroy even with tyndallization. Today the strategies center more around the application of heat as a part of a program of chemical treatment. In these cases, the heat increases the rate of chemical reactions as well as traumatizing and killing the biological entities.

Disruption by freezing is another technique which is now being evaluated and patented processes for applying these principles have already been sought. The principal theoretical consideration is that when a biofilm is frozen, the bound water within the ECPS will gradually freeze into ice crystals. Once these ice crystals are formed, the whole of the slime is structurally disrupted so that when the system is thawed out, the biomass completely shears away from the surfaces. Once this biomass is removed by pumping, the resident surfaces appear pristine and hold the promise of the well recovery being to original specifications.

Methods presently under development call for the biozones to be frozen by the application of coolant (e.g., cold ethylene glycol, solid carbon dioxide) directly down the water well or through injection at a number of sites around the biofouled well. Concerns relate to the potential problems arising from the physical distortions that may accompany the freezing and cause excessive pressures which could damage the structures in the water well. The methodology remains in the prototype demonstration phase.

pH manipulations. Biofilms are vulnerable to radical shifts in the pH of the environment. Normally, most biofouled groundwater systems have pH (water) ranging from as low as 5.5 to 9.5. The pH of the incumbent biomass within the attached sessile phases will tend to buffer the pH to within a range from 6.8 to 8.6. This effect is microbially manipulated to optimize the conditions for satisfactory growth, metabolism and survival. Applications of acids to the biofouled regions around a water will cause very rapid downward shifts in the pH values depending upon the strength and volume of the acid applied. Generally, the key factor which would influence survival would be the pH falling down from a value of 5.5 through 3.5 as the acidity increased. Beyond a pH value of 3.5 down to 1.5, this decline in pH is likely to impact more severely upon polymeric structures within the biofilm causing hydrolysis (breakdown in the structural integrity) with some total digestion (to simple soluble organic molecules). However, the pH gradient is established across the biofilms (i.e., lowest pH at the water:slime interface and the highest pH in the deepest strata of the biofilms). The maximal impact of the acidified conditions therefore has a more intensive and immediate impact on the surface layers of biofilm within the slime with disruption, sloughing and dispersion occurring with a high degree of lethality. Deeper biofilm layers are subjected to a much more gradual increase in the acidity (i.e. fall in pH). This may provide sufficient time to allow some protective response reactions to occur. These responses can include: (1) releases of some of the bound water leading to a shrinkage in volume; (2) intrinsic buffering reactants functioning to neutralize and reverse the falling pH along the diffusing pH gradient; (3) the incumbent microorganisms initiate stress survival responses (e.g., generation of endospores, retrenchment into ultramicrobacterial states, reduced metabolic activity states); and (4) enact a sloughing event wherein the incumbents within the sessile suspended particulate masses can move on water flows to new habitats. The act of sloughing (4) is most likely to lead to the destruction of the organisms during the transport phase with the water flow due to the ready penetration of the particulate structures by the acids used in the treatment. The most likely survival in an acidization event are the traumatized microorganisms still incumbent within the compressed polymeric structures supporting the residual stressed slimes.

Acidization can prove to be an effective controlling agent against biofouling only in conditions where the pH gradient rapidly extends through the various strata of biofilms within the slime. This rapid acidic extension throughout the strata has to cause the total disruption of the slime before stress initiated responses can enact. Once buffering (2) of the impact of the pH drop is initiated and the incumbent microflora enters a protective survival state (3), there is an increasing likelihood that there will be a sufficient critical mass of survivors so that a fresh biofouling could be quickly initiated.

Another major concern is that a minimum amount of the disrupted (e.g., sloughed particulates) and the hydrolyzed/digested biofilms be left in the treated zones of the biofouled porous media. Such "debris" should be pumped out as completely as possible from the treated formations. This would reduce the total mass of "residual nutrients" that could again be utilized by any of the survivors of the treatment through mechanisms (2) and (3) described above.

Acidization offers a relatively simple approach to the control of clogging events around a water well where these are dominated by dead inorganic (e.g., carbonate salt structures, silts and clays) and organics (e.g., organic rich muds). In these events, there would be relatively little reactive response to the acidic conditions generated and the processes of hydrolysis and digestion would allow the residual nuisance material to be pumped from the well. Where there is a plugging event with a significant amount of biological activity, the probability of the acidic manipulation becomes reduced. However, acidization can form a major part of a synergistic (blended) treatment of a biofouled water well.

Ionizing radiation

When Roentgen discovered X-rays in 1895, it was quickly discovered that these ionizing radiations could kill microorganisms. Today, ionizing radiation is divided into two principal types. These are corpuscular and electromagnetic types. Corpuscular radiations are due to subatomic particles of various types which transfer their kinetic energy to anything they strike. The energy released is relatable to the speed at which the particle is travelling. Electromagnetic radiation is produced as waves by high voltage equipment. The shorter the wavelength the higher the energy content. Both X-rays and gamma radiation are electromagnetic waves produced from high voltage equipment and the atomic nuclei respectively. X-rays range in wavelength from 10^{-6} to 10^{-10} CM with frequency cycles of from 10^{17} to 10^{21} frequency cycles per sec. while gamma rays have a shorter wavelength (10^{-11} CM) and higher frequency (10^{21} frequencies per sec). Gamma radiation is now used in the raduridization of a variety of foods (such as lamb, fish and spices) to significantly reduce the risks of microbial spoilage and food infection. Gamma rays are generated from such radioactive isotopes as cobalt-60 and cesium-137. There is a potential to control biofouling at a fixed site through the application of gamma radiations. It is measured in rads (0.01 Joule) as radiant energy. Dosages are measured in kilorads (Krad) for 1,000 rads and megarads (Mrad) for 1,000,000 rads. Sterilization is usually considered to be achieved by 4.5 Mrad (equal to 45 kJ/kg). Research has focused on the use of low dosages in the Krad range being delivered continuously to the potential site of biofouling. This is done by placing a cobalt-60 source of adequate strength close to the area of concern.

In the food industry, the typical dosage for radurization is between 100 and 1,000 Krad. By practice, this has been found to reduce the microbial population between 90 and 99%. Lethal dosages (Krad) for a range of species are: humans, 0.6 to 1.0; insects, 25 to 100; bacteria and fungi, 100 to 1,000; bacterial endospores, 1,000 to 5,000; and viruses, 3,000 to 5,000. In the water industry, the use of gamma radiation has been developed in the eastern regions of Germany and Czechoslovakia for the control of biofouled water wells. There are a number of patents which have been issued relating to the concepts and processes associated with this technology. The advantages of this type of treatment are claimed to be the: (1) suppression of biofouling within the zone of influence; (2) restricted ability of microorganisms to attach within the radiated zone; and (3) blocking of the transmission of viable hygiene-risk microorganisms (e.g., coliform bacteria) through the treated zone. Disadvantages relate to the (1) cost of installing a gamma ray emitter close to the perceived site of biofouling; (2) perceived risk that there would be a generation of induced radioactivity and the formation of unacceptable byproducts as a result of radiation; (3) difficulty and costs involved in the maintenance of the installation and its eventual decommissioning; and (4) risk of unacceptable chemical releases.

Blended treatment technologies. All of the various technologies discussed above do offer advantages and disadvantages in the control of microbial activities within the surrounding porous media and in the well's water column. In the last decade, more attention has been directed to the development of more complex treatment strategies. These involve bringing together different treatment technologies in a manner that would maximize the likelihood of success. In designing the precise nature of the blending of these strategies, the objective to be achieved is to remove the biofouling from around the water well to a sufficient extent that a prolonged period of satisfactory performance can be expected. Such blended treatments will be, by nature, more expensive than the use of a single-treatment approach. It is therefore critical that the blended treatment is applied in a customized manner to the biofouled installation not only to maximize the potential for recovery but also to generate a monitoring procedure which would allow future treatments (single or blended) to be initiated appropriately.

One example of the use of a blended treatment strategy is the patented blended chemical heat treatment (BCHT). This method involves a triple-phased strategy of chemical treatment to disrupt the biofouling while, at the same time, heat is applied to increase the effectiveness of the chemical treatments. The BCHT process operates in three distinct phases of application (Figure 26) which involves the shock, disruption and dispersion of the incumbent biofilms within the volume to be treated. Each phase is discussed below.

Shock phase (BCHT 1). This initial phase is designed to maximize the degree of trauma within the stratified biofilms forming the fouling phenomenon. A triple-treatment approach is used where there is the simultaneous application of heat, disinfectants and surfactants. The heat is applied to take the temperature up into the lethal range within the treatment zone. Terminal temperatures are dependent on the ability of the well installation to resist the combination of heat and chemical effects. Terminal temperatures in the shock phase therefore range from 65°C to 95°C. The disinfectant should penetrate throughout the biofilm to, along with the thermal impact, cause the death of most of the microorganisms. A surfactant is employed to increase the rate of diffusion of the disinfectant into the biofilms.

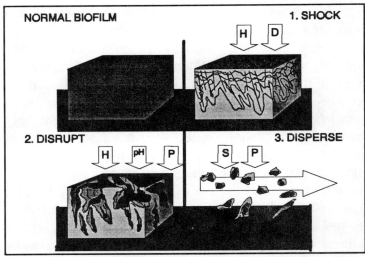

Figure 26. *The patented Blended Chemical Heat Treatment (BCHT) involves the use of heat (H), acid (pH), disinfectants (D) penetrants (P) and surging (S) to rehabilitate a biofouled well through three phases.*

Disrupt phase (BCHT 2). A pH shift (usually downwards into the range of 1.5 to 3.5) initiates phase 2 and a second surfactant is applied which functions efficiently to disrupt the traumatized biofilms generated by phase 1. Heat continues to be applied. During this phase the effect of these conditions is that the biofilm structures are now disrupted into hydrolyzing structures and much of the incumbent microbial population is destroyed.

Disperse phase (BCHT 3). By this stage in the BCHT treatment, most of the biofilm structures within the treated zones are severely disrupted while any biofilms immediately outside of this zone are severely traumatized. The objective of phase 3 is to disperse the residual materials still within the zone of treatment. As a prime concern, this objective must include the removal of these materials from the treated zone in a manner that will preclude the possibility of this material re-entering these zones to support a re-infestation. Phase 3 does not employ heat and so the system is allowed to cool down while a surging process is used to keep the treated materials in suspension until pumped out the well. There is therefore a sequential surge - pump to waste scenario until the water being pumped out exhibits no physical or chemical evidence of the treatment effects (e.g., low particulates, ambient temperature, turbidity and pH, and "normal" chemical constituents). A third surfactant may also be applied during this phase to ensure that the dispersed materials remain in suspension. It should be remembered that much of bioaccumulated materials (e.g., iron and manganese salts) will also be pumped out of the well in the phase 3 pump phase. Disposal of this waste may be subject to local and national regulatory control.

Ongoing monitoring of the product water after a blended treatment should include a determination of the residual microbial surviving population which may originate in the mildly disrupted zone outside from the intensively treated zone. If there is a significant amount of debris remaining in this treated zone, there could be rapid recolonization by survivors. The rate at which microbial aggressivity occurs may be seen to increase over time in the product water. For example, decreases in the *dd* of BART™ testing over a series of monthly testing periods can be used to trigger a treatment scenario. In Waverley, Tennessee this approach has been used since 1988 to control a biofouling event in some municipal production water wells which was controlled by BCHT. Using the SRB-BART™ tests on a monthly basis, the *dd* were between 12 and 14. Periodically there would be shifts down to 8 to 10 and a shock disinfection caused the *dd* to return to 12 to 14. There were minor releases of manganese (the major bioaccumulate of concern) during the period of reduced *dd*. In 1990, there was a sudden reduction in the *dd* from 12 to 6 and on down to 4. This preceded a major release of manganese in the product water. A repeat of the BCHT treatment caused the removal of this excess manganese bioaccumulation and the system again stabilized with *dd* values again in the 12 to 14 range.

It is important in any water treatment to include an ongoing monitoring where there is a biofouling event suspected. Such monitoring should be used to ensure that an adequate level of treatment is applied to suppress the biofouling and produce a product water of acceptable quality.

7

Sampling Methodologies

Sampling a water system to investigate a potential microbiological problem is not as simple as just turning on a faucet and filling any container with the first water to gush out. There are several basic factors to be considered which would affect the value of any data obtained from this water sample. Influencing factors could include contamination by microbes growing within the faucet itself; the microbial activities which would have occurred in the pipe to the faucet at temperatures which may perhaps be very different from that of the well; any viable microbial cells present in the sampling container; casual infection of the water with airborne or skinborne microorganisms arising from a dirty room or poor operator techniques; erratic microbial composition in the main water line being sampled because of spurious microbial events such as the initial turbulence associated with the startup of pumping. Each of these factors and any combination can cause serious changes in the reliability of the final results obtained. This would be equally valid not only from direct testing methods, but also from indirect testing techniques. It is therefore essential to follow sampling procedures which will minimize errors.

151

CONTROLLING FACTORS IN SAMPLING

Sampling procedures must always be adjusted to the local circumstances relating to the particular water source. These adjustments can be addressed by determining the likely interference levels that could be experienced and adjust the sampling procedures accordingly. These adjustments are discussed for each of the potential interference factors listed in the following paragraphs.

The point at which samples are to be taken should always be carefully selected since this choice may influence the types of microorganisms that are likely to be recovered with the water sample. In all cases this selected sampling point should be as close as possible to the actual body of water that is a concern. Whether the sampling point is a hose extension from a gate valve or a direct flow from a faucet, there will be a distinct possibility that a unique microbial flora will be growing in association with the particular installation (e.g., the hose, the faucet). The types of activities that can be expected to occur may also include attached growths over the surfaces that are permanently wet with a film of water. This environment will most likely support aerobic microorganisms which will grow within a biofilm to form a slime. If the device is leaking slightly there would be an ongoing source of water and nutrients so that the generating slime may grow out over the discharge port and even splash down to form additional slime growths in the drainage channels. These bacteria would be growing at the ambient (room) temperatures which may be considerably different from that of the main body of water to be sampled. If an accurate sample is to be taken of the targeted water then steps have to be taken to exclude these "contaminating" bacteria which are growing around the opening to the faucet or valve. Where this occurs (for example, in the hose attached to the sampling device), remedial steps have to be taken to divorce these factors from the sample being taken. Two therapies are commonly used for reducing these interferences. They are: (1) a prolonged running of water through the sampling system in order to "flush" out any of the detachable microorganisms from the port and/or hose so that it is mostly the incumbent organisms in the main body of water being sampled that are present; and (2) heating the sampling port to a sufficiently high temperature to destroy these nuisance contaminants.

Flushing the sampling port is difficult because the rate at which microbial incursions into the flowing water occurs from the surfaces of the sampling port and connecting line to the main water body cannot be predicted accurately. The microflora occupying the connecting line also has the potential to be very different from the organisms occupying the main water body due principally to the lack of flow which will probably be experienced in this line and the potential thermal gradient which may generate along the line between the temperature of the water body and that of the sampling site. This gradient would become particularly significant where the temperature differential (difference between temperature of the main body water and that of the environment within which the sampling port was situated) exceeded 10°C or there was a 5°C shift across the 15°C or 45°C thresholds. These thresholds form barriers to the continued activities of some groups of microorganisms. For example, if water was being pumped from a groundwater source at 8°C and the sampling port was at 22°C, a temperature gradient of 14°C would be created along the connecting line. Additionally, the water would have crossed 15°C and a biased population that may be expected to occur due to the shifting in the dominant temperature-related groups of organisms.

Where flushing is being used to obtain a microbiologically satisfactory sample of the main body of water, there is always a risk of some interferences from the flora originating in the connecting line and sampling port. To reduce these risks associated with the flushing method of sampling, the water should be run for long enough to generate a stable temperature in the product water which should be equivalent to the temperature of the body of water being sampled. When the temperature of the flowing water to be sampled is the same as the targeted water body, some flushing activity may be considered to have occurred. However, there should be an extended period for running the water after the stable target water temperature has been reached before actually taking the sample. The selection of the length of this extended period of flow prior to sampling is difficult to standardize due to the ongoing risks of contamination. In general, a minimum of ten minutes of flow beyond the acquisition of a stable water temperature would minimize these risks. For heavily biofouled water systems, a series of samples may be taken at 5, 10 minutes, 1 and 2 hours after temperature stability has been observed and the water continues to flow.

Temperature measurement may be achieved by a number of techniques. At the simplest level, the temperature of the flowing water may be tested by the prudent placing of the hand within the flow periodically. Where the mainline water is markedly cooler, this technique may function satisfactorily. For a more precise monitoring, a bulb thermometer (with a minimal range of 0°C to 60°C) or an electronic temperature recording probe can be placed directly into the flowing water. The temperature can then be used to determine the start of the stable temperature acquisition period.

Microbiological contamination risks can be further reduced by destroying the viable (living) microbial flora growing in the sampling port itself. This can be done by lightly flaming the port with a propane torch until the port begins to steam where the port is recognized to be made of heat resistant materials. This should be done with extreme caution respecting the ability of the materials to withstand such a heating process. For example, a gate valve constructed of brass would be reasonably suitable for such a treatment while a plastic valve assembly would not. An alternative heating process would be to pour very hot or boiling water over the sampling port for long enough to ensure that the temperature of the port assembly rose until it became too hot to be comfortably touched. The net effect of such a heating of the port is to destroy many of the micro-organisms in the water and/or slime formations within the device. Subsequent flushing may remove a lot of the detached detritus resulting from the heating process. An extra two minutes of flushing should be added to compensate for this when heating the port as an additional safeguard.

METHODOLOGIES

On some occasions, it is possible to access the sampling site directly by threading a flexible tube down to the targeted area in the well. Such tubes commonly are flexible walled plastic materials such polyethylene or polyurethane. The flexibility allows the water to be pumped from the site by peristaltic or flap valve pumping provided that the head generated is not too large. In many cases such tube insertions remain on site for the period of the project and may even become a permanent part of the installation. In all cases, the insertion holding patterns will lead to a potential for extensive biofouling of the line so that when water is pumping through the line there will be alterations to the microbial loading of the water which will reduce the value of the information obtained. Avoidance of this influence can be accomplished by the insertion of a sterile tube which is plugged at the entrance port with a plug which may be blown out at the point where sampling is to occur. Repetition of this technique would allow a higher quality of water sample (freer from intrinsic biofouling) to be obtained but would be expensive to perform. Application of disinfectants such as sodium hypochlorite can be used to reduce the covert contamination of the sampling tube itself but may cause secondary impacts by suppressing the natural microflora being pumped through the line.

To reduce the concerns about anomalous data due to the biofouling of the water sample during passage through the sampling tube, one of the simplest approaches would be to prolong the pumping period prior to the actual act of withdrawing a sample. This prolonged pumping should allow (minimally) ten volumes of water to be pumped when the water is crystal clear or up to fifty volumes of water where the water is distinctly turbid. A volume is defined as the internal volume of the sampling tube from the entrance port to the sampling port. Where there is a sequence of sampling which is monitoring a treatment being performed associated with the targeted water, a minimum of 1.5 volumes should be pumped to discharge prior to each sampling.

The simplest method for sampling water is to collect it from a stream of water being discharged from a pipe, or from a reservoir receiving such waters. Such sampling may be influenced in terms of accuracy by the environmental conditions prevailing around the port at that time. For example, if there was a strong wind, some of the airborne dust could enter the water sample. To minimize these concerns, water samples may be drawn directly through the wall of the plastic sample tube from the midpoint of the water flow. The sequence for sampling by this manner is to use the double-ended needle, holder and sterile evacuated sampling tube method. This technique is commonly known in the health sciences field as an acceptable manner for

withdrawing blood samples and is referred to as the vacutainer (brand name, Becton Dickinson, Rutherford, New
Jersey) system. The steps to perform this type of sampling
are:

1. Clean off the surface of the tube into which the sample
   needle is to be injected with a clean paper towel.
2. Moisten a fresh clean paper towel with either an alcohol
   or 30% peroxide solution and rub over the surface of the
   tube so cleaned. Allow the tube to dry.
3. In a clean place, remove the cap from the single sample
   needle and screw the needle into the holder so that one
   needle now extends upwards into the chamber. Flip off the
   cap from the other end of the single sample needle to
   expose a larger needle. Carefully insert this larger needle
   through the treated wall of the sampling tube and position
   the needle at the midpoint of the flow. Note that the tube
   should have a sufficient clarity to allow this positioning to
   be viewed.
4. Once the needle is in the correct sampling position, lower
   the evacuated sampling tube (vacutainer) tube, rubber cap
   first, down into the holder containing the smaller exposed
   needle. When the cap is touching the needle press the tube
   hard onto the needle.
5. Water will be sucked up into the sterile vacutainer tube by
   the vacuum action within the tube causing it to fill to
   about 70% of capacity with water. When the charged
   vacutainer tube is withdrawn, the cap will seal and a water
   sample would have been taken with a minimum of external influence.

One limitation to this technique is that the water sample
is essentially pulled into a vacuum and there could be
generated a shortage of oxygen which may be stressful for
any strictly aerobic organisms. It should be noted that the
vacutainer system does involve the use of surgically sharp
needles and care should be exercised in the use and disposal

of this equipment.

Single samples of water may be obtained from vertical water columns such as exist in many wells by the use of bailers. There are a wide variety of items of equipment which involve the principle of entrapping a volume of water at a predetermined depth. These devices may passage through the water relatively freely and then seal a volume of water when remotely triggered by a mechanical or electrical signal. The passage of the units through the water, even when the units have been presterilized, can become fouled with attaching microorganisms prior to the triggering of closure. In some cases the mechanical disturbance created by the movement of the device through the water column may further exacerbate these interferences.

One counter measure to reduce interference risks when using a bailer is to send the sample counter down in a sterile bell filled with sterile (filtered) compressed air which prevents (by positive pressure) water from being admitted to the bell (Figure 27). The air supplied by a compressed air line from the surface can be controlled and the pressure recorded at the surface. When the bell has reached the desired depth, the air pressure is released causing water to flood into the bell and overflow into the sterile vertical incumbent sampling vessel which would be open on the upper side. To recover the water sample, the flooded water is blown out of the bell with a surge of sterile compressed air and the bell returned to the surface. Here, the sample tube is removed within the stream of compressed sterile air and capped. This system of sample collection minimizes the potential for an accidental contamination of the sample tube but, in practice, is time consuming and expensive since the bell assembly would need to be sterilized prior to each reuse.

The next stage in the sampling is to ensure that the water samples obtained do not become contaminated casually with organisms coming from the environment or from the

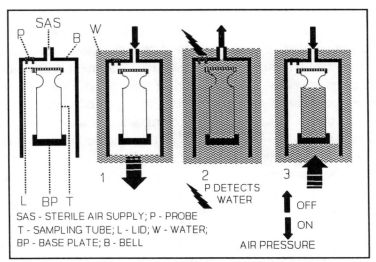

Figure 27. *Diagrammatic presentation of the bell sampling method for taking a water sample from a set depth in tube (T) by positioning the bell (B) and reducing air pressure (arrows).*

operator of the sampling procedures. Care and attention has to be directed to ensure that no dust or dirt is likely to be admitted into the sampling device, that no operator directly handles the inside surfaces of the container or any part likely to be in direct contact with the water sample. The sampling container should be sterile and sealable. Sterile containers by definition must contain no living (viable) cells and should also not contain any chemicals likely to inhibit microbial survival. As a result of these constraints, most sampling containers are made either of glass or nontoxic plastic materials. In general, the glass containers have a screw-down metal cap and are sterilized by autoclaving (15 p.s.i. steam @ 121°C for a minimum of 20 minutes). These containers are often reused many times after repeated cleaning and sterilization cycles. Plastic sampling containers are presently usually considered to be disposable and therefore are used only once. High-density polyethylene can, in some grades, be autoclavable and hence reusable. The common 125 mL coliform bottle is a common choice as a container. The disposable plasticware is often sterilized by either gamma irradiation or ethylene oxide. Some plastics are, however, resistant to autoclaving (e.g., polyurethanes) and so can be cleaned and resterilized for repeated use.

To take a sample from a flowing water stream, the container should be first labelled with sample origin, date, projected time for taking the sample and the tests that should be performed on the sample. Once this is done and the time is approaching to take the sample, the cap is carefully unscrewed and laid down on a clean surface without turning the lid over. The clean surface is important since any dust and dirt on the surface may contaminate the inside of the cap (and hence the water sample). If there are doubts as to the cleanliness of the surfaces then these surfaces may be swabbed down with a 10% dilution of a domestic bleach solution which contains 5 to 6% sodium hypochlorite. Allow the surface to dry before beginning the sampling procedures.

At the time of uncapping, the sampling bottle would have a sterile interior which is now being exposed to the air. Do not touch the inside surfaces, handle the container by gripping the lower part of the bottle and avoid any skin contact with the upper part of the container. When taking the water sample, gradually place the open container vertically into the water flow and allow the water stream to fill the bottle. Once the bottle is two-thirds to three-quarters filled, move the bottle sideways out of the water stream and quickly cap and seal the bottle. Screw the cap down tightly.

In most cases, the sampling container will be rigid-walled and designed for the task but in some circumstances, such as when an emergency occurs or when there is a limited amount of storage space, these containers may not be available. There does exist a variety of flexible walled containers commonly manufactured from polyethylene film and shaped into a bag with a twist tie attachable to seal the container once filled. Such bags are sterilized in advance and provide a light weight method for carrying the uncharged vessels. The plastic is usually of a sufficient gauge that there is a very small probability of a rupture and leakage of the sample. As a last resort, where no sterile containers are available and a water sample has to be obtained in order to pursue a microbiological line of the investigation, the rolls of plastic bags retailed in supermarkets can provide a "quick and dirty" alternative. It is recommended that where this method is pursued, every effort be made to select a bag from within the roll (where there is a higher probability of the inside of the bag being sterile). Once the sample has been taken (fill to no more than 60% of the bag's maximum volume), the bag should be sealed tightly with a twist tie and the whole bagged sample sealed again within a second plastic bag to reduce the risk of leakage. Details of the sample source and time should be written on the outer bag using a suitable permanent marker. It should be remembered that these domestic use plastic bags are not sterile and hence should only be used as a last resort or where examination is going to be limited to nonculturing activity (e.g., microscopy, staining, chemistry).

TRANSPORTATION

The next stage is to transport the sample to the testing facility where the microbiological tests would be performed if not conducted on site. One major concern in transportation is trying to ensure that the microorganisms remain viable (alive) and active without multiplication until at least the testing procedures have been performed. The very act of taking the sample would have exposed the microorganisms to sudden changes in pressures, temperatures with variable exposures to light. Any combination of these factors may cause some of the microorganisms to go into various stages of trauma, which may become a lethal event for some of the organisms. Care must therefore be exercised to ensure that the water sample has been taken with due consideration to the origin of the water being investigated.

Storage of the water sample during transit to the testing station should be in the dark away from any possible solar heating of the sample containers and also protected from freezing. When such transit times are relatively short (i.e., less than six hours) the water should be maintained at as close to the original ambient temperature of the water as possible. These conditions may be achieved by storing the samples in an insulated container and packing in such a way that the intrinsic heat in the water stabilizes the temperature of all of the water samples to a common median level. If the water has been collected from cold water sources with a median temperature of below 8°C, it would be advisable to add some ice packs to restrict any temperature elevation. If the water has been collected from sources at above 30°C, then the intrinsic heat in the water samples should be sufficient to maintain the population without additional trauma. In some events, shifts in the microflora size and composition can occur within three hours.

In many cases, six hours may not be a sufficiently long time to allow the transit of samples to be completed. The general consensus in such cases is to lower the water sample temperature to less than 10°C in order to restrict the amount of growth and deleterious interactions (e.g., competition) between the intrinsic species present in the sample. Without temperature control and with the additional oxygen that is often present, the water samples may be subjected to growth blooms and species dominance shifts which would detract from the value of the microbiological results obtained. For example, the changes in environmental conditions could cause a suppression in the coliform bacteria while stimulating radical increases in the numbers of pseudomonads.

The standard method employed to control these post-sampling shifts in the incumbent microflora is to provide a cold temperature shock to the floristic system. Such control shocks are implemented where samples are to be in storage and transit for periods of between six and twenty four hours. To apply the control shock, the water samples are packed around bags of ice in an insulated cooler for storage in the dark preferably in a cool place (to lengthen the time of the cooling effect). The objective is to bring the water down quickly (at -2 to -8°C/hr) to within the range of 1 to 4°C. At these temperatures the incumbent microflora undergo a number of reactions. There will be a reduced activity to a basic survival metabolic mode, some species will become less competitive with the rest of the flora, and interact to a lesser extent with the available nutrients and other chemicals in the water. Some species may seek such protective econiches as may be provided through attachment and the production of additional polymeric slime material. When these survival modes are initiated, less floristic changes occur in the water sample. At the same time, however, the component organisms are now in a minor state of trauma. Prolongation of the low temperature storage for periods of longer than twenty-four hours is generally considered to cause the traumatic conditions to become so severe that the nature of the incumbent microflora will be irreversibly affected. Even with an acceptable storage period of, for example, twenty hours there may be a prolongation of recovery times needed for some of the organisms. This trauma may cause their presence to remain undetected in subsequent test routines. This is particularly a problem where growth or metabolic function is monitored over a very restricted time frame. For example, a two day spread plate evaluation for heterotrophic bacteria at 25°C by counting the colony forming units may generate only a small percentage of the viable bacteria able to grow on the medium used because many traumatized organisms are still recovering from the cold shock trauma.

In some events it remains impracticable to return the water samples for laboratory examination within twenty four hours of the samples having been taken. In these cases involving prolonged storage, any data so obtained is likely to be influenced by the differential effects of the cold shock trauma on the incumbent microflora. Interpretation of data obtained using these samples must therefore be more restrictive and relate to a comparison of the microfloral composition of the various samples taken concurrently from the same or similar water source(s).

Perhaps the conclusion to these limitations imposed by the need to transport conserved water samples to a laboratory for examination is that the ideal condition would be to eliminate the storage period entirely and conduct all microbiological testing immediately in the field at the sampling site. Direct technologies which can be employed at site include the use of microscopy, of prepared agar plates and dip paddles, BART™ biodetectors, and dehydrogenase activity test systems.

8

Procedures to Estimate the Degree of Biofouling

INTRODUCTION TO BAQC

Biofouling around a water well can involve a number of events which can range from changes in the taste and odor generated in the water by microorganisms, through to the generation of a cloudiness or color resulting directly and/or indirectly from microbial activities through to a direct interference with the passage of water into the well and through the pumping process to the header transmission lines. These latter events can involve plugging where the capacity of the well is observed to be significantly affected. This can mean that the well or well pump is no longer able to pump water at the same flow rates due to an occlusion (plugging of the pore spaces) leading into the well or through the well screen or at the pump intake. Well impairment may also be observed by increases in the drawdown of water during the initial startup of the pumping procedure. If there is an increasing resistance to flow of water into the well, this is directly reflected in an increasing rate of drawdown of water in the column head which compensates for this increasing resistance to flow resulting in plugging. Pump clogging is indicated by a reduced output at any drawdown.

Biofouling can therefore take a number of forms in the pumped water and relate to different events. Generally speaking, the sites for microbial growth associated with such biofouling events would be connected to the water column itself and objects in it such as pumps, any well screen through which water is infiltrating into the column, the gravel pack and/or other packing media that is used around the screen and finally, the media in the aquifer itself. Consequent biofouling will vary to some extent with the site. For example, in the well water column head, there are less surface areas to which microorganisms can attach. Consequently, the microbial activity that may be visualized as occurring within this area would occur either in a suspended (planktonic) form, or as attached biofilm growths. This latter growth would be on surfaces presented by such piping, casing and screen material as may be exposed to the inside of the well. The direct TV logging of a well or the examination of pulled equipment will reveal these types of growths as discoloration, tuberculations, encrustations, slimy overgrowths, protruding slime masses and floating particulates within the water or upon surfaces. During the period of time that a well is quiescent, a stratification of these suspended microbial growths will frequently occur so that higher populations of microorganisms can be observed in a series of descending strata with the highest density of microorganisms being recovered: (1) at the base of the well and (2) in zones where the maximum amount of water enters the well through the screens.

The other zones around the water well will present relatively large surfaces on which microorganisms can attach, colonize, and grow through to forming a biofilm which may eventually become confluent and occlusive (i.e., not allow water to pass through the porous media). The well screen (when present) by its very nature may provide a barrier to the entry of the solid particles into the water column. At the same time, these slots allow water to pass through. This action involves a constriction at the points where water is able to pass through the filter screen. Such zones often form sites for intensive microbial attachment and fouling. Supporting fouling at these sites is enhanced by the rapid changes in the turbulence of the water as it enters the column. Water flow may become turbulent at this point but becomes a more passive laminar flow further from the well. Increases in the concentration of oxygen in the water may stimulate the aerobic microorganisms. Rapid passaging of suspended and dissolved nutrients moving into the water column head from the groundwater source may also stimulate microbial activities.

Frequently, observation of fouled slots reveal heavy microbial growth around those slots which are actively allowing water to passage into the well. As the biofilm (slime) grows across the slot, the available area for water to enter the well may become significantly reduced and consequently, the flow of water may first of all be diverted to unfouled sections of the screen. Eventually fouling will reduce the total flow into the well water column. Outside of the screen in the packing material and in the fractured and porous aquifer media, biofouling can occur on the many and often varied surfaces presented.

It has been observed that such biofouling tends to maximally occur in the zones where the oxygen concentration is extremely low (i.e., the zone between where oxidation or reduction processes may alternately dominate). Where this reduction-oxidation fringe moves closer and further away from the well, there may tend to be a broader based biofouling over that whole zone. Some methods for controlling the biofouling of wells utilize this phenomenon of creating an optimal growth at the oxidation-reduction border by extending the size of the oxidation zone out into the groundwater system by the direct injection of aerated (oxygenated) water at convenient points around the production well. One such method is the "Vyredox" patented in situ iron removal water well treatment system.

The nature of biofouling may therefore include both suspended planktonic particulates in the water well column and attached growing biofilms occurring on surfaces within and around the well itself. While the suspended materials may readily be pumped from the well and identified to be indicating that microbial growths and a biofouling event are occurring, the occurrence of a biofilm based biofouling is more difficult to predict.

Using TV logging, the occurrence of distinctive slimes and encrustation formations downhole can be observed as an indication that a biofouling is occurring within the well. It is, however, impossible for such observations to determine the degree of biofouling occurring beyond the direct view of the camera. If the reduction-oxidation fringe is supporting a biofouling three feet (one meter) beyond the well screen, such observations would have no significance. Observations must involve an understanding of the mechanisms that function in the growth, maturation and disintegration of biofilms deep-seated in the aquifer media around the well.

The beginning of biofouling is initiated by bacteria attaching to the surface. Once attached, the bacteria reproduce and then colonize the surface by either gradually spreading over it, rolling over and reattaching or hopping from one place to another, reproducing at each site. When the bacterial cells have begun to develop, they form a protective envelope of slime (polymers) which retains water, nutrients and protects the cells from adverse environmental conditions. As colonization proceeds over the surfaces, so the different species of bacteria interact. The competition that results causes the resultant biofilm to contain a consortium of different strains of bacteria which are now able to co-exist for mutual benefit. Often as the slime thickens, the bacteria requiring oxygen dominate in the upper strata of the slime while the bacteria able to grow in the absence of oxygen (anaerobic) form the deeper strata.

Initially, the biofilm thickens very rapidly as the microorganisms maximize the volume of occupancy and compete for resident roles in the developing biofilm. This causes the slime to grow very rapidly and occupy a large percentage (up to as much as 50%) of the pore spaces. As the consortia within the biofilm stablizes, this volume shrinks again, allowing much more rapid passage of water. A secondary maturation of the biofilm occurs after the completion of the shrinkage (compression phase). This secondary expansion is somewhat erratic and involves two principal phenomenon. First, the biofilm will increase measurably in volume and subsequent to or during this event, there will be an increase in the resistance to water flow. This increasing resistance to water flow may be due to the slime throwing out polymeric materials into the water as a series of strands which then cause a considerable resistance to water flow. The effect of this on the product water is for that water to contain particles of sheared biofilm and flow more slowly. These sheared particles will also contain some of the incumbent bacteria, particularly from the upper strata of the biofilm. Since this event is erratic, the occurrence of these sheared particulates may be variable over the course of pumping. The product water (postdiluvial) being pumped from the well may contain a mixture of resuspended planktonic and sheared biofilm particulates.

To determine whether a biofouling event is occurring, one or both types of these particulates should be observable in the product water. There is an enhanced possibility of recovering such material from a water that has been immediately pumped from a water well held in a quiescent state. Here, the water has not been forceably pumped through the water well system to cause an ongoing dilution of any shearing and sedimentation effect.

Once the various biofilms become confluent within the porous media around a well, there is likely to be a serious occlusion (plugging) of the porous media and water is no longer able to reach the well in quantities close to the original observed capacities. Where this occurs, biofouling may also be suspected and proven by the presence of higher populations of microorganisms in the pumped water.

There remains, however, the serious question of the early prediction of biofouling during the phases when the biofilm is just beginning to form and occupy a significant volume within these pore spaces, particularly at the oxidation-reduction zone fringe. The observation of a biofouling event must therefore involve an expeditious monitoring of the postdiluvial water at the time when biofouling is most likely to be observable. This would be after a period of quiescence (when the well has not been pumped, the biofilms are in stress). During this period, the planktonic microorganisms settle while the sessile forms go into stress (i.e., change in environmental conditions). These sessile forms tend to shear from the biofilms and enter the void spaces.

There are three time series sampling procedures for the biofouling assessment quality control (BAQC) of the water well recommended to determine a biofouling event. These include a rapid test (Double Two) and a longer, more accurate procedure (Triple Three), and the comprehensive BAQC-16.

THE QUICK DOUBLE TWO TEST TO DETERMINE THE OCCURRENCE OF BIOFOULING IN A WATER WELL

This sampling technique has been designed to be quickly performed and give some indication of whether a biofouling event is occurring. It involves three phases: (1) shut well down for a period of two hours; (2) reactivate pumping and sample at two minutes into the pumping cycle; (3) leave the pump running and sample again at two hours. Two water samples are obtained at two minutes and two hours into the pumping cycle. These samples should be taken with reference to the standard sampling procedures and may be subject to a variety of testing which is relevant to parameters associated with biofouling. These include: particle counts, particulate loading, turbidity, bacterial activity (assessed by such technologies as the BART™) and the direct enumeration of the number of bacteria utilizing either the standard laboratory techniques of extinction dilution spreadplate analysis or the agar dip paddles. The object of the Double Two is to provide a sufficient length of time when the well is not pumping whereby some sedimentation of the particulates in the water column would have occurred.

Concurrently the biofilm which may be developing around the water well would be free from the turbulence associated with pumping and therefore more likely to undergo a limited expansion with the possible generation of finger-like polymeric structures moving out into the water. Two hours would be a very minimal time to allow these events to occur. At the same time, however, it would allow a complete evaluation of a well to be accomplished in half a day. When the pump is started up after the two hours of quiescence, there would be a resuspension of the settling particulates and a shearing of any newly forming processes arising from the stabilized biofilm matrix which had been generated under continuous pumping. These two components (resuspending and shearing particulates) would be pumped from the well. After two minutes of pumping, the water would be dominated by resuspended particulates and any sheared material which had occurred close to or within the water well itself. After two hours of pumping, the resuspended particulates would all have been now virtually removed by the pumping process and the shearing of biofilm close to the well would have been complete.

As a consequence, the two hour sample may be expected to resemble more closely the background particulate loading in the water. To confirm this, a control sample is taken immediately prior to switching the pump off and having the well enter the quiescent phase (before quiescence, BQ). This control sample would represent the background amount of activity which is generated in the well in terms of particulate and biological loading during an ongoing active pumping process. It would be preferable for the pump to have been activated for at least 24 hours prior to the initiation of the Double Two sampling procedure. Where biofouling is not occurring, there should be no difference in the biological population, activity levels, the total volume of particulates or turbidity in all three of the samples. Where biofouling has occurred, an increase in the population, activity, particle loading and/or turbidity may be expected to have occurred. The interpretation procedure is given in the second section below.

THE TRIPLE THREE BAQC TEST TO DETERMINE THE OCCURRENCE OF BIOFOULING IN A WATER WELL

The Triple Three scenario is similar to the Double Two but involves a longer period of quiescence and an extended period of sampling beyond that recommended for the Double Two. While the Double Two sampling can be achieved in four hours, the Triple Three takes three and half days to complete and therefore is something that would best be performed by the operators of a water treatment distribution system rather than consultants. The technology involves maintaining the well in a quiescent state for twelve hours instead of two hours prior to sampling. This extended period is designed to give a greater time factor for the sedimentation of planktonic particles and for the extension of the biofilm mass into more vulnerable finger-like processes entering the water phase.

Four samples are taken in this procedure. These are: (1) prior to switching the pump off at the beginning of the quiescent period; (2) three minutes after the initiation of pumping; (3) three hours; and (4) three days after the initiation of a continuous pumping cycle for the well. The same parameters may be monitored as for the Double Two sampling (see section below) and biofouling may be recognized by a radical diversion of the three-minute sample from the presample (1) and these three-hour postquiescent sampled (3). If the three-hour pumped sample (3) shows considerably higher biological and particulate activity than the prequiescent sample (1), this could be interpreted to indicate that a deep-seated biofouling was occurring within the well system and extended further out into the aquifer.

SUMMARY

It can be seen from the above discussion that the form of a biofouling event is relatively unpredictable and forms in ways that can be unique to each water well so affected. There are, however, a number of common events which can be associated with biofouling and these may be used to predict the likelihood of such an event occurring in a given water well.

The Double Two and Triple Three scenarios represent some of the simplest approaches to a determination of the size of a biofouling and more intensive monitoring can actually yield data which, upon interpretation, can be used to calculate the relative extension of the biofouling out beyond the water well into the aquifer media itself. In determining whether a biofouling event has occurred using the Double Two or the Triple Three test scenario, it is important to utilize more than one method for potentially recognizing that biofouling has occurred (Figure 28). In both the Double Two and Triple Three test scenario, the water sample is taken before the well is turned off (made quiescent). The before quiescence (BQ) sample should be taken at a predetermined point when a dormant period for the well can be tolerated by the user.

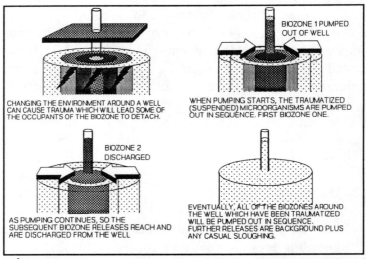

Figure 28. *Sequence (top,left-right;bottom,left-right) of BAQC disruption (trauma created by changing operating conditions) and pump out sequentially of each biozone affected.*

If the water production demand is excessively high, the Double Two test scenario should be used, which requires a quiescent time of only two hours. It should be remembered that the BQ sampling should be performed after the well has been operating continuously for a period of preferably greater than twenty-four hours and, before that, been in relatively continuous use (i.e., greater than 60% of the time). During the period of quiescence, the suspended particulates within the well water column will settle and the biofilms around the well will be free of excessive turbulence of the type associated with the pumping activities (Figure 29).

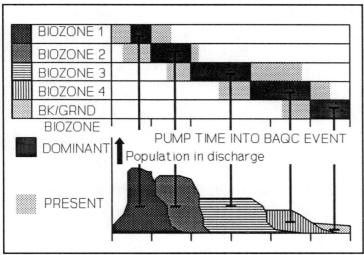

Figure 29. *Schematic sequence of biozone discharges (shaded, upper linked to graphic lower) over the time period of pumping out the traumatized biozone material during BAQC (horizontal axis).*

The samples drawn at the recommencement of pumping will include some additional biological loadings where a biofouling event has occurred and introduced products (for example, increased levels of suspended and sheared particulates) into the water. Tests can be undertaken using the various standard methodologies or the cruder semi-qualitative/semi-quantitative BART™ biodetector which can be employed in a more remote location.

DIAGNOSTIC LEVELS FOR SIGNIFICANCE

Double Two Test Scenario

Interpretation using the BART™ dd data

Where a range of different BART™ tests were applied to water samples taken from a water well before the quiescent phase (BQ), after two minutes of pumping (D1) and again after two hours of pumping (D2), it becomes possible to assess the biofouling risk using the table below which shows the **minimum acceptable** dd values for D1 and D2 which can occur after different control dd values have been obtained from the BQ sample. Where dd values obtained for the D1 and D2 samples are **lower** than the values shown in columns two and three below, the probability of biofouling can be considered to be more **significant.**

Before Quiescence (BQ); 2 mins into pumping (D1); 2 hours into pumping (D2)

BQ	D1	D2
*3	2	3
*4	3	4
*5	3	5
6	4	6
7	5	6
8	6	7
9	6	8
10	7	8
11	8	8
12	9	9
13	9	10
14	10	11
15	10	12

*wells would automatically be considered biofouled, values given are dd values for RX1 using the BART™ biodetectors.

Scan along the first column for the *dd* value obtained in the BQ sample and determine whether the *dd* values obtained for the D1 and D2 values are equal to or less than the values shown in the BQ column. If these *dd* values are lower, then a greater level of biological activity has been observed and biofouling can be suspected. The severity of the biofouling may be crudely gauged using the following formula:

$$B_x = dd(D1) / dd(BQ)$$
$$B_y = dd(D2) / dd(BQ)$$
$$B_z = (2 - (B_x + B_y)) * 100$$

where *dd*(BQ), *dd*(D1) and *dd*(D2) are the days of delay to reactivity for the samples taken before quiescence (BQ), and at the first and second post-quiescent sampling (P1 and P2 respectively). Factorial shifts in the microbial aggressivity as a result of quiescence are calculated as B_x for the D1 and B_y for the D2 sampling. The biofouling severity index (B_z) is calculated from the factorial shifts and expressed in the scale where a positive value indicates that a biofouling event is occurring and the severity of the event may be expressed semi-qualitatively using the scale of +1 to +200. A negative B_z would support the hypotheses that no biofouling was evident using this procedure. Severity may be expressed using the following chart:

B_z value range	Prognosis
Less than -1	Biofouling not evident
0.0 to +25	Minor biofouling projected
+26 to +50	Probable biofouling
+51 to +100	Significant biofouling
+101 to +150	Major biofouling
> than +150	Extreme biofouling

It is recommended that the *dd* data used be obtained from the particular BART™ system causing the shortest *dd* values in BQ test. Where the *dd* values so obtained are shorter than 5.0 in the BQ samples, it may be construed that the well is already biofouled to a significant extent. Further reductions in the *dd* in the D1 and D2 samples confirm this event.

Bacterial Populations by Spreadplate Enumeration

	(cfu/mL)	
BQ	D1	D2
* < 100,000	> 250,000	> 150,000
* < 50,000	> 125,000	> 75,000
* < 10,000	> 50,000	> 20,000
* < 5,000	> 25,000	> 15,000
< 1,000	> 5,000	> 4,000
< 500	> 2,500	> 2,000
< 100	> 1,000	> 500
< 50	> 500	> 250
0	> 500	> 250

Determine which row is appropriate to the BQ population of the selected group of bacteria (left-hand column). If the populations recorded after quiescence exceed the numbers shown in the right-hand two columns (D1 and D2), then biofouling can be suspected. If only one (usually the D1) shows elevated populations, the biofouling may be considered of a relatively minor significance. Note that where the BQ population exceeds 5,000 cfu/mL, an intense biofouling may be occurring within the water well system (*) and remedial measures may need to be taken.

An example of this would be if the results obtained were (cfu/mL): BQ, 184; D1, 5,200; and D2, 870. Here, a scanning of the row where the BQ was <500 (but greater than 100) would show that both the D1 and D2 results exceed the acceptable maxima and so biofouling may be considered to be occurring.

Where a number of different groups of bacteria have been enumerated, compare each group separately since not all may be involved in the biofouling. Groups showing elevated populations in the D1 and D2 samples are more likely to be associated with a biofouling event. If the population is focused around the D1 sample, the fouling is probably close to the water well itself and/or the transmission system from the well to the sampling port (in cases where a down hole sample could not be taken). In cases where the specific group of bacteria has a heightened population around the D2 sample, it may be construed that the biofouling by that particular group is more probably deep seated within the porous media surrounding the well. A more precise determination of this type of fouling would require the comprehensive BAQC survey involving sixteen timed samplings after quiescence.

Turbidity (NTU)

Where there is a significant (greater than 2 NTU units) increase over the BQ value, biofouling can be suspected if there are also increases in the rate of biological activity (recorded by the BART™ biodetectors) or population size by spreadplate enumeration.

Pigmented Particulates (% pp)

Very often there are pigmented particulates suspended in the water samples where there has been an ongoing bio-accumulation of oxidized iron within the fouled zone. This gives the particulates an orange, red-to-brown color. When 100 mL of a water sample rich in these pigmented particulates is filtered through a 0.45 micron membrane filter, there are two effects. Firstly, the white color of the membrane filter will become overlayered with the pigmented particles. The extent of this discoloration will be relatable to the loading of the water with pigmented particles (i.e., the darker brown the filter becomes, the greater the loading borne by the water). Secondly, the particulate material will begin to plug the 0.45 micron pores in the membrane filter which would extend the filtering time for the 100 mL water sample. This time extension will reflect not just the pigmented particles in the water but the total particulate loading which is plugging the pores of the filter. The former technique is therefore more appropriate to estimation of pigmented particle loading in water.

Estimation of the particle loading may be assessed either visually (crude) or by the use of a reflected light photometer. In both cases, a comparison is made between a PANTONE white color paper/uncoated (as 0%) and the PANTONE 469 brown Letracolor color paper (as 100%). The technique employed involves the membrane filtration of 100 mL of the sample water through a standard 47 mm diameter 0.45 micron plain (not gridded) membrane filter used for the coliform test. Once the water has filtered through the membrane, the filter is removed and air dried. It should be noted that the filter, while still wet, cannot be conveniently read due to the reflectivity of the water film on the membrane. Where the filter needs to be dried quickly, it is recommended that the filter be held by forceps (at the edge) in a stream of hot air from a hair drier. Do not hold the filter too close to the drier since there would be a risk of the hot air causing color changes in the pigmented particles and heating up the forceps to a point where it is uncomfortable to hold. Once the filter is dried it may be kept in a cool dry dark place prior to reading and as a reference.

Visual estimation of the level of pigmented particulates is undertaken by comparing the color on the filter to within the PANTONE range from white (as 0%) to brown 469 (as

100%) and estimating its density within that percentile scale. More accurate estimations can be obtained by using a reflected light photometer. One such type of reflected light photometer (model IRP, Droycon Bioconcepts Inc., Regina, Canada) uses a 600 nanometer adjustable 4 mm light beam which is reflected at an angle of 80° from the horizontal to a vertically set photocell positioned 150 mm above the filter. Since the pigmented particulate (*pp*) entrapped on the filter absorbs light, the lower the amount of light received by the photocell, then the greater the amount of *pp* entrapped. This is measured as the inverse of the light received from the white control filter. The scale is set by scaling the unit to 100% using the PANTONE 469 paper. Thus the higher the percentile reading, the greater the presence of *pp* and the greater the potential for an iron rich biofouling to be occurring.

Using the *pp* values obtained in a Double Two test, the following observations may be made.

B.Q.	D1	D2
*<30%	>50%	>40%
*<25%	>35%	>30%
*<20%	>30%	>25%
*<15%	>25%	>20%
<10%	>20%	>15%
< 5%	>10%	> 7%
< 1%	>10%	> 5%
<-1%	> 5%	> 5%

Percentile *dd* values reflect the level of pigmented particles in 100 mL of water. Scan the rows for the BQ data appropriate to the results obtained. If the D1 and D2 data exceeds the values shown, biofouling may be considered to have occurred. If the excessive value occurs only in the D1 data, the biofouling with *pp* may be limited to suspended particles and/or shearing biofilm inside or close to the water well. Confirm that a biofouling event has occurred by evaluation of the biological data (as significant biological presence should be noted). If no biological activity has been observed, a chemical oxidation-encrustation event may have caused the higher incidence of *pp* observed and a biological event may not been significantly involved.

Direct Membrane Filtration

Where there just needs to be a rough evaluation of the risk of biofouling, this can be done by examining the rate at which 100 mL of water filters through a 47 mm 0.45 micron filter. This can be done by recording the time taken (filtration time, ft) for the 100 mL water sample to pass through the membrane filter. The greater the density of suspended solids in the water, the higher the probability that the membrane filter flow will become impaired and the longer the time needed for the 100 mL sample to pass through the filter. A 100 mL particle-free water can usually be filtered through a membrane filter in less than twenty seconds. Where there is a considerable quantity of suspended solids in the water, these particles will impede the passage of water through the filter membrane and a much longer ft will be recorded. In badly fouled situations, ft times of as long as eight minutes have been recorded. To use this ft technique, the following protocol should be used:

1. Set up the membrane filtration apparatus with a fresh 47 mm 0.45 micron membrane filter in place. Note that it does not need to be sterile.

2. Conduct a control trial using 100 mL of a crystal clear distilled water and time the period from pouring the water into the filter to the loss of a continuous water film over the membrane. This time period would form the control ft.

3. Clean the membrane filtration equipment and insert another fresh membrane; it is not necessary to sterilize the unit.

4. Time the passage of 100 mL of the BQ water sample through the filter and record as the BQ-ft. Repeat stages 3 and 4 for the P1 and P2 water samples.

5. In turn, deduct the control value ft from the BQft, D1ft and D2ft to obtain the extended ft generated by any particulates fouling the water. These would be recorded as ft_q, ft_{d1} and ft_{d2}, respectively. For example, the ft_q would be calculated using the following equations:

$$ft_q = BQft - control\ ft$$
$$ft_{d2} = D2ft - BQft$$

This extended ft data may be used to indicate the potential for particulate fouling in the water sample following the criteria listed below.

Where the ft_q is >20, the water may be considered to be permanently fouled with particulate material which may be composed of silt, sloughed slime, clays, sessile suspended particulates and planktonic microorganisms. Other biological, physical and chemical tests would need to be employed to determine the nature of this fouling.

Where the ft_{d1} and the ft_{d2} exceed the ft_q by at least twenty seconds, there can be considered to be a likely biofouling event occurring around the water well. If only the ft_{d1} value exceeds the ft_q value, then the fouling may be considered limited close to the well screen and within the well itself.

Where the ft_{d1} and/or the ft_{d2} exceed the ft_q by at least sixty seconds, the fouling may be considered very severe. It should be remembered that these prolongations of the ft are as a result of the plugging of the membrane filter with suspended particulate material from the 100 mL water sample. This fouling could have been due to various causes (e.g., silt, clays, bioslimes) and further testing (microbiological for the verification of various bacterial groups, and chemical for the concentrations of iron, silt and clays) would need to be performed to verify the cause.

Triple Three Test Scenario

Where there is more time to perform the testing but still a restricted budget and staffing, the **triple three** test scenario can be employed. Here, a water sample is taken before the period of quiescence (BQ), and after quiescence three more samples are taken at three minutes (T1), three hours (T2) and three days (T3) after the commencement of continuous pumping. The extension of the sampling period to three days (i.e., T3) has been designed to allow for the recording of any larger biofouling events. If the readings obtained from the T3 are similar to the BQ data, the biofouling may be considered finite and restricted (i.e., the biofouling would be clearly restricted to a distinct zone around the water well). If the T3 data exhibits considerably higher readings than the control (BQ) and similar to the T2 reading, the biofouling can be considered "infinite", (i.e., not restricted to the zone of influence of the water well) and will require the more intensive BAQC-16 investigation to more precisely define the scale of the biofouling.

Similar tables to the double two scenario are displayed below for comparative purposes. The critical difference is the T3 data which, when it is significantly above the BQ values, could indicate that a very deep-seated biofouling event was occurring.

Interpretation Using the BART™ *dd* Data
Before Period after initiation of pumping

Quiescence	3 mins	3 hrs	3 days
BQ	T1	T2	T3
*3	<2	<2	<2
*4	<3	<3	<3
*5	<3	<4	<4
6	<4	<5	<5
7	<5	<5	<6
8	<5	<5	<6
9	<6	<7	<8
10 to 14	<6	<7	<8

Determine the *dd* for the BQ reading (left hand column) and scan across row appropriate to the data. If the *dd* vales for the T1, T2 and T3 readings are equal to or are less than the *dd* values shown, biofouling may be suspected. Note that biofouling will be occurring where BQ values of less than 6 are obtained (marked with an *). Where a lower *dd* value is obtained for only one of the three T samples, biofouling may be restricted to a relatively narrow zone around the water well. If only the T1 shows a lower *dd*, the fouling would most likely be close to or inside the well column. If it is the T2 or T3 which shows the lower *dd* value, the fouled zone may be expected to be further away from the well.

Bacterial Populations by Spreadplate Enumeration (cfu/mL)
Before Period after initiation of pumping

Quiescence	3 mins	3 hrs	3 days
BQ	T1	T2	T3
*< 100,000	>250,000	>150,000	>125,000
*< 50,000	>125,000	> 75,000	> 60,000
*< 10,000	> 50,000	> 20,000	> 15,000
< 5,000	> 25,000	> 15,000	> 10,000
< 1,000	> 5,000	> 2,500	> 1,500

<	500	>	2,500	>	1,000	>	750
<	100	>	1,000	>	250	>	150

Determine which row is appropriate to the BQ data obtained and examine T1, T2 and T3 data to determine whether these values exceed the stipulated levels. The extent of the biofouling can be projected where the data exceeds the limit shown. Another method for the interpretation of the bacteriological data where the samples exceeded the limit is given in the interpretation table below (samples designation given for those higher than the limit) set for that value (e.g., >5,000 cfu/mL). It must be remembered that these interpretations are indicators and should not be treated as absolute indicaters:

T1, T2 and T3 - heavily biofouled
T1 only   - biofouling close to well
T1 and T2   - biofouling extends out from well
T2 and T3   - an intense biofouling is occurring around the well but not in the well
T3   - biofouling seated well away from the well

Pigmented Particulates (%pp)
Before Period after initiation of pumping

Quiescence	3 mins	3 hrs	3 days
BQ	T1	T2	T3
*<30%	>50%	>40%	>35%
*<25%	>35%	>30%	>30%
*<20%	>0%	>25%	>25%
*<15%	>25%	>20%	>20%
<10%	>20%	>15%	>15%
< 5%	>10%	> 7%	> 7%

Determine which row is appropriate to the BQ result obtained and examine the T1, T2 and T3 data to project the likelihood of a biofouling event occurring. Where the BQ data exceeds 10% (*), there may be an intrinsic biofouling event occurring within the water well. The location and scale of the fouling may be projected using the interpretation table shown under b immediately above this section.

BAQC - Sixteen

A Structured Determination of the Size of Biozones around a Biofouled Water Well

The Double Two and the Triple Three techniques can be used to determine the probability of an active biofouling occurring within and around a water well. The methodology involves the shutdown of the water well for a period of time (quiescence) during which the biofilms forming the fouling will become stressed. One result of this trauma is that some of the biofilm will slough (or detach) and enter the water. When the pump is turned on again, the pumped flow will include a magnified amount of the particulate matter originating from the fouling. The greater the releases and the longer these releases continue may be used to "guestimate" the size of the zone of biofouling but there is an inadequate sampling to make any quantitative estimate of the scale and dimension of the biological intrusions into the water well and surrounding aquifer.

The BAQC-16 involves a more comprehensive sampling and analytical procedures which allow a superior estimation of the scale of the biofouling. Two major objectives can be achieved using the BAQC-16:

1. Estimation of the dimensions (volume occupied and diameter) and relative locations of any biozones formed around the water well.

2. Determination of the various microbial consortia which form the identified biozones within the biofouled zone.

To achieve these objectives, a series of water samples (sixteen) are taken in sequence from the water well once pumping is restarted after an extended period of quiescence (e.g., seven days). The sequence of sampling involves sampling the water at intervals which should entrap water being received by the well from a increasing zone of influence. Unless there are local conditions which dictate an alternative approach, the standard sampling procedure would involve eighteen samples, two of which would be the control samples taken at one hour (BQ1) and one minute (BQ2) before the pump is turned off to initiate the quiescent period. It should be noted that the well should preferably have been in production on a continuous basis for the previous seven days and on at least 70% of pumping capacity for the month before then. These two samples (BQ1 and BQ2) should represent the characteristics of the water well under conditions of constant pumping with a minimum amount of sloughing (limited to randomized and local events). After the period of quiescence (minimum, 7 days) sampling should proceed from the moment that pumping is initiated as the D series (Figure 30). Recommended sampling times are:

D1 - 2 minutes	D2 - 5 minutes	D3 - 10 minutes
D4 - 20 minutes	D5 - 30 minutes	D6 - 1 hours
D7 - 2 hours	D8 - 3 hours	D9 - 4 hours
D10 - 6 hours	D11 - 9 hours	D12 - 12 hours
D13 - 1 day	D14 - 2 days	D15 - 3 days
D16 - 4 days		

As the D series samples are being taken, the water is reflecting the shifts in the characteristics of the water as it "flushes" out from a larger and larger zone of influence around the water well. Characteristics recorded from these samples can be critically analyzed to aid in the determination of the size, location and composition of the biozones. Where there is found to be no increase in the loading of the D series water when compared to the BQ series, it can be construed that there is no evidence for a significant biofouling.

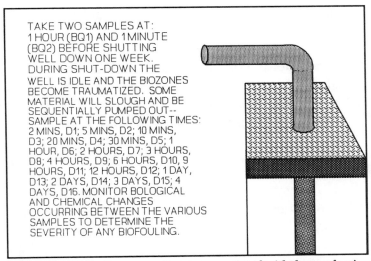

Figure 30. *Sequence of events associated with the conducting of a BAQC-16 investigation for the presence and size of a biofouled zone around a water well.*

Characteristics which can be included in the BAQC-16 incorporate:

Laser Particle Sizing (MPA), can be used to determine total suspended solids (ppm), mean particle size (micron), particle volume distribution and total number of particles as a part of microbial particulate assessment **(MPA)**. It allows the total volume of the particulates to be projected based upon the mean particle volume released over the period of enhanced releases of particulates (PERP).

Period of enhanced releases of particulates (PERP), the time period in which the D series samples exhibited significantly greater loadings than that recovered from the BQ series. A significant increase may be considered to be >50% higher data than that recovered for the mean of the two BQ series samples. PERP may extend through minutes (minor biofouling) to hours (moderate biofouling) or as long as days (severe biofouling).

Total Iron (ppm), where a water has become fouled by bacterial activity, one effect is that the iron in the causal water is bioaccumulated in the biozones fouling the well. This means that the postdiluvial water being pumped from the well may not include this "biofiltered" iron. In the BAQC-16 procedure, the biofilms within the various biozones may become severely traumatized. In this state, the initiation of pumping will cause initial very high concentration of iron to be pumped from the well to give very much above normal total iron values. The length of time during which these high iron values are present in the pumped water after quiescence can be used to project the size of the zone around the water well where excessive bioaccumulation of iron had occurred. This may be found in some circumstances to coincide with higher aggressivity (i.e., shorter *dd*) in the IRB population.

Total Manganese (ppm), like iron, manganese may also become bioaccumulated within the biozones around a water well and be released at least in part when postquiescent pumping is initiated. On some occasions the manganese may be accumulated preferentially to the iron so that the bulk of the manganese may be deposited in biozones further out in the formation.

Iron:Manganese Ratio (Fe:Mn), the tendency for the iron to be bioaccumulated in the biofouling closer to the well than the manganese (Figure 31) may be observed in the shifting Fe:Mn ratio which may be seen occurring in samples D1 through to D12. In the early water samples where there has been severe IRB biofouling the Fe:Mn ratio can reach a range very favorable to the Fe (e.g., 100:1). However, as the sampling program continues, the Fe:Mn ratio may fall until close to parity (e.g., 2:1) or even a manganese dominance (e.g., 1:10). Other metals may also be bioaccumulated as various oxides, hydroxides and other salts at various points within the various biozones (Figure 31).

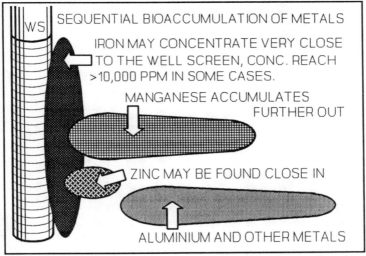

SEQUENTIAL BIOACCUMULATION OF METALS

IRON MAY CONCENTRATE VERY CLOSE TO THE WELL SCREEN, CONC. REACH >10,000 PPM IN SOME CASES.

MANGANESE ACCUMULATES FURTHER OUT

ZINC MAY BE FOUND CLOSE IN

ALUMINIUM AND OTHER METALS

Figure 31. *Diagrammatic presentation of the regions within the biological interface (see Figure 18) beyond the well screen (WS) where various metallic compounds are bioaccumulated in a sequential manner.*

Concentrations can sometimes be found of such elements as aluminium, chromium and cadmium which can sometimes be recorded where the groundwater system is located close to ore bodies, natural or industrial leachages and/or tailings disposal sites. Dates from the discharge of disrupted biofim from wells treated by BCHT have revealed that the various metallic cations concentrate in different ratios within the various biozones around a well. Iron tends to accumulate by MIA close to the well, while manganese accumulates further out. Periodic releases from these D zones would shift the Fe:Mn ratio very significantly in the postdiluvial water (Figure 32).

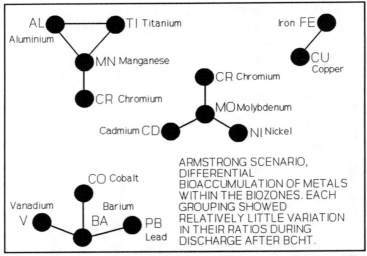

Figure 32. *Interrelationship between various metallic elements observed in the discharged material obtained from the Armstrong well after BCHT treatment. Ratio variation < 40% of mean value.*

Total Organic Carbon (TOC, ppm), where there is a relatively young biofouling it is probable that the basic chemical structures within the suspended particulates will include a high percentile of organic carbon. This organic carbon would be employed for three primary functions: (1) the generation of ECPS to entrap water around the cells, provide protection and act as a storage reservoir; (2) as a major component within the cells occupying the particulate material; and (3) as stored organic material within the particulate structures. Some additional organic carbon may be in the soluble form dissolved in the carrier water. The TOC therefore gives some indication of the potential for a biologically based fouling through the levels of organic carbon recovered.

Incumbency Strategies to Determine Biofouling, the nature of the sampling procedure would dictate a high probability that much of the "recoverables" analyzed for (see lists immediately above) would have originated directly or indirectly from the various biozones surrounding the well. If this is the case then the incumbency of these observed recoverables within the determined particulate volume may aid in the establishment of the scale of biofouling.

Incumbency may be determined by taking the data (e.g., cfu/mL) and relating this to the volume of particulates which are being expressed as TSS (ppm). Essentially, the parameters are measurements in viable units per 10^{-6} m^3 (cfu/mL), grams per m^3 (ppm, w/v), or milliliters per m^3 (ppm, v/v). When the assumption is made that all of the viable units will reside in particulate structures then it becomes possible to project the incumbency density of these viable units per unit volume of particulate structure. The assumption is reasonable since the laser-driven particle size determination will function down to 0.5 microns, below the normal lower limit for microbial cell lengths. Incumbency for the viable units (**IVU**) can therefore be established as the number of viable units occupying a fixed volume of particulate material. This can be done using the following equation:

$$IVU \quad = \quad VU \quad / \quad TSS$$

where IVU is the calculated density of viable units per 10^{-12} m^3, VU represents the cfu/mL and TSS the particle volume in ppm. For example, where a spreadplate enumeration was performed for **HB** and gave a population 5,700 cfu/mL and the water sample was determined to have a TSS of 12.7 ppm, the IVU would be 449 vu/10^{-12} m^3 of particulate material. In the case of determining the incumbency of a specific substance (or group of substances) which had been determined in ppm (w/v) then the incumbency for the gravimetric units (**IGU**) using the following equation:

$$IGU = (PPM / TSS)$$

where IGU is the calculated weight of the substance in micrograms per 10^{-12} m^3, PPM is the gravimetric analysis in grams per m^3 and the TSS is the particulate volume in millilitres per m^3. For example, where there was a total iron (**TI**) of 12.6 ppm and the TSS was 4.6 then the IGU could be calculated as being 2.74 micrograms/10^{-12} m^3.

The IVU and IGU values thus give a relationship within a fixed volume for the levels of incumbency both of various microorganisms and chemical substances respectively. In a very active biofouling the GVU values may reach as high as 20,000 or more vu/10^{-12} m^3 while in a relatively passive system the GVU may be very low (e.g., 20 to 40 vu/10^{-12} m^3). Where the IGU is found to increase in the P series of samples the bioaccumulation of that substance may be considered to have occurred further out in the formation.

Establishing the Theoretical Position of the Biozones, the data from the D series samples should be first examined to determine if these are significantly higher in particulates, viable entities and specific chemicals than the BQ series. If there are greater loadings, attention can be directed towards establishing the relative positions of any distinctive biozones associable with the biofouling of the water well. A number of basic premises have to be considered to differentiate one biozone from another:

1. that the bulk of the sloughed product (i.e. particulates) from a biozone arrives at and is pumped from the well in sequence with materials arriving from other biozones. In other words, the initial sloughed product would come from the closest biozone followed by the material from the next (concentric) biozone and so on until the material from the outermost biozone has been voided (pumped out). From that time on, the pumped water would not contain any major amount of sloughed material from a particular biozone but may contain occasional non-specific and minor releases from the various biozones.
2. each biozone would have some unique characteristics which would clearly differentiate it from the biozones on either side. These differences could include a different bacterial group dominance, changes in the IVU and/or IGU, shifts in the chemical concentrations.

All of the data from a BAQC-16 determination needs to be gathered together in a spreadsheet format with the columns (vertical) set to each of the BQ and then P series (eighteen in total) while the rows (horizontal) will be for each of the tests that have been performed routinely on all of the samples. For the BART™ tests, both the *dd* and the **RX** should be included on separate rows for each test type. IGU and **IVU** values should also be included where these can be calculated. Note also the flow rate (**Q**) of water from the water well during both the BQ and the P series testing. The **Q** should remain the same for both series. From the data, each biozone may be recognizable through a common pattern of data events which separates these samples from the preceding (inner biozone) and "postcedent" (outer biozone). Once the sample sequences have been established for each of the biozones (e.g., biozone 1, 0 to 5 minutes; biozone 2, 10 to 30 minutes; biozone 3, 1 to 4 hours; and biozone 4, 6 to 12 hours) it becomes possible to extrapolate the volume of the biozone (**VB**) and the theoretical diameter of both the inner and outer diameters of the zone (**ID** and **OD** respectively). All three computations can be performed either manually or using the WELLRADII software program. This allows a theoretical extrapolation of the (cylindrical) size of the biozone and fouled volumes. From this information, a treatment strategy may be developed.

To calculate the **VB** for a given biozone, the following parameters have to be included: **Q**, time to completion of the preceding biozone (**T$_c$**), time to completion of flow from the biozone being calculated (**T$_f$**) and the porosity of the materials immediately surrounding the well. These porosities may include a fixed diameter of highly porous fill (**P$_h$**) such as gravel pack which would be surrounded by a lower porosity (**P$_l$**) natural aquifer material. Porosities estimated as a fraction of the total volume (e.g., a porosity of 0.4 would mean that 40% of the total volume was porous and could be occupied by water).

Volumes can be calculated more conveniently by projecting that the biozone abuts the precedent (inner biozone) and extends outwards to the outer encompassing biozone. The volume of the biozone (**VB**) would be comprised of that flow which would be pumped from the water well through from the time at which the precedent biozone terminated (**T$_{pc}$**) to the termination time (**T$_{tb}$**) for the observation of the "marker" presences of the biozone under determination. The **VB** may be calculated using the formula given below:

$$
\begin{aligned}
\mathbf{V_p} &= \mathbf{Q} \times \mathbf{T_{pc}} \\
\mathbf{V_t} &= \mathbf{Q} \times \mathbf{T_{tb}} \\
\mathbf{VB} &= (\mathbf{V_t} - \mathbf{V_p})
\end{aligned}
$$

where V_p is the volume preceding the initial releases from the biofilm and V_t is the total volume released by pumping to the termination of the observation of the "markers" specific for the identified biozone. The volume of the biozone (VB) is the difference between the V_t and the V_p volumes.

Once the volume(s) of the various biozones have been calculated, the sum of these volumes represents the minimum total volume of biofouled water around the water well that will need to be treated if the event is to be suppressed (by either the temporary or radical suppressive strategies). While it is essential to understand the volume of biofouled zones around a well, it may also be "administratively" important to make the users aware of the distance that such fouling extends around the well. Should the option be selected to abandon the water as being too biofouled to allow a satisfactory recovery, a replacement water well may be planned and installed within the same groundwater system. It is naturally highly desirable that the new installation does not intrude directly into the fouled biozones around the existing well and that a safety factor be built in to reduce this risk. Clearly, there would be a high risk of biofouling where the installation did actually penetrate through one of the concentric biozones. While there can be no absolute confidence that the new installation would not become biofouled, the greater the physical distance is from the biofouled well the lower the probability that fouling will occur. As an initial strategy of concern to minimize the risk, a new installation should be installed as far away from the fouled water well as possible and a minimum distance (D_{min}) of the total diameter (D_{zbz}) of the sum of biozone widths around the well may be calculated using the following equations (1) and (2):

(1) D_{zbz} = $2 \times R_{max}$

where R_{max} is the radius of the largest defined biozone from the midpoint of the water well, and:

$$(2) \quad D_{min} \quad = \quad 5 \times D_{zbz}$$

The biozone radius for this calculation can be determined using the WELLRADII described in Appendix Two.

Once the D_{min} has been calculated, this distance can be used as a guide to the distance that a new water well may be installed with a reduced risk of biofouling. However, it should be remembered that the whole groundwater system may be vulnerable to such foulings, in which case, the additional distance may simply serve to delay the onset of symptoms of the infestation.

In some water wells, the biofouling may be close to the well installation itself but appear as a series of pulses as each lateral layer of the biofilm sloughs off. The net effect would be similar to the inward movement of materials from the various concentric biozones established around the water well. In the event of a lateral sloughing, the above projections would exaggerate the size of the affected zone. The result of this would be to be build in larger "safety factors" than are used for the concentric biozone form of infestation.

While much of the discussion relates to biofouled water wells that are producing, it must be remembered that biofouling also poses a major risk to pressure relief, monitoring and idle water wells that are out of production. For these wells which are "passive" for much of the time, the application of the BAQC scenario utilizing a period of quiescence is clearly not appropriate. A strategy has to be incorporated to "force" the detachment of some of the sessile attached microflora so that these presences can be observed. This may be achieved by the pumping of the well which stresses the incumbent microflora living in a relatively passive environment. For this approach, the BQ series would be samples taken from the passive water column, while the P series would be taken through the startup of the pumping period. If prolonged pumping is not practical, then a mild application of chemicals such as a disinfectant or heat may produce the same level of trauma. After the traumatized (and killed) cells have been pumped, there should follow a period of magnified releases of viable cells as a result of the trauma created.

9

Monitoring Methodologies for MIF Events

INTRODUCTION

Perhaps the major challenge in attempting to monitor the occurrence of a microbially induced fouling (MIF) in a groundwater situation is the remoteness of the site from the sampling point. Rarely can satisfactory samples be taken at the actual site of the MIF because of the porous structures within which it is located. Downstream sampling (usually at the well head or beyond in a storage or distribution system) is therefore a convenient substitution. While convenient, such downstream samples may become subjected to anomalous events due to the addition of microbial entities functioning downstream from the site, and the loss of some of the fouling organisms due to such events as predation, competition or alternate site colonization. Such remote sampling may therefore not accurately reflect the activity which is occurring in a focused manner at the MIF site.

Given the potential errors which are intrinsic in remote monitoring, it remains essential that the methodologies employed allow an appropriate assessment of the MIF. Such interpretations can occur over a range of domains from the simple presence/absence, through to approximate or accurate determinations of the numbers (quantitative) or types (qualitative) of microorganisms functioning both at the MIF site and along the conduit through which the water passes to the sampling site. In addition to an evaluation of the biomass in terms of numbers and variety, newer methodologies are focusing more and more on the level of activity being performed by these incumbent organisms. A very small but active (aggressive) microflora may have a greater impact on the dynamics of a particular MIF site than a large, but passive, microflora. These impacts may range through shifting hydraulic transmissivity (e.g., plugging and/or facilitative flows) to fluctuating bioaccumulation, biodegradation and sloughed releases of specific chemicals[50]. Such complexities in the chemistry and hydrological characterization of the product (sampled) water through this range of interactive events renders scientific interpretation that much more challenging where an MIF event is occurring. Indeed, the conjecture can be proposed that, where there is a randomized significant departure within an individual or a group of characteristics, an MIF event should be considered as a possible significant factor.

The development of an approach to determine the presence and significance of an MIF event can include a number of strategies. In summary, these approaches may be summarized into two pathways: (1) indirect; and (2) direct.

Indirect pathways of determination do not involve the direct determination of the biological entities which may be involved but, rather, examine shifts in the physical and/or chemical characteristics of the product (postdiluvial) water which could be attributable to any form of biological activity.

Direct determination of the involvement of biological entities within a groundwater system can involve direct microscopy (possibly coupled with staining techniques);

selective culturing and subsequent enumeration; chemical determination of specific compounds directly associated with microbial cellular activities; or an evaluation of the potential aggressivity of the entities when presented with standardized environmental conditions[5,8,66,73].

All of the methodologies being described are dependent in part, for their validity, on an acceptable system being employed for the sampling. Given the frequent remote location of the actual MIF focus site, it is very important that the sampling procedure takes into account the need to sample or "acquire" water from the suspected fouled site. There is, by the nature of the conditions prevalent, always going to be a level of uncertainty that the sample obtained reflects accurately the microflora occurring at the site of concern. One net outcome of this uncertainty is that the absence of microorganisms from a given sample does not necessarily exclude these organisms from being involved in the biofouling event under investigation. It could simply mean that none of these entities happened to be present in the water at the time of sampling. Conversely, a positive indication for the presence of a given microorganism could be a reflection of a "contamination" event in which the identified organism entered the water from a site other than the fouled zone (e.g., downstream econiche point of use device[33,71]). The degree of uncertainty can be reduced by a more frequent and diligently controlled sampling program.

The "double-two", "triple-three" and BAQC sixteen water well sampling programs already described are designed to allow an increasing level of certainty to be obtained by structuring the sampling program to increase the potential for recovering organisms from a focused site of the fouling. Of these three techniques, the double-two is the least sophisticated and the BAQC sixteen is the most comprehensive. Where water samples have now been taken, the next concern is the nature of the examinations that will then be conducted.

In the above premises it is assumed that the major investigation may be via the water samples obtained. Another investigative route is by downhole observation. Visual logging for the presence of biological growths within the water column or attached to the casing and well screens can be achieved using a submersible TV camera. Where these growths are attached, they become readily recognized as large, sometimes mucoid or plate-like structures which often extend out into the water column. If the growths are dispersed in the water, these may appear as visible, often "fuzzy" suspended particles or as a general cloudiness. The color of the growths may reflect whether there has been iron and/or manganese accumulation within the growths[81]. Deeper orange-red to brown colors may reflect an iron dominance while blackening may indicate either a high manganese content or the presence of large concentrations of metallic sulfides. The absence of any such growths does not automatically mean that there has been no biofouling of the water well and the surrounding aquifer systems. It could be that the site of the biofouling may be more distantly located from the well and therefore not directly observable. Where such a remote site undergoes a sloughing event, some suspended particulates may be observed in the water.

Some attention has been paid to the determination of microbial biofouling by the admission of devices on, or within which attachment and growth would be encouraged to occur. Such systems may include glass, plastic or metal pristine sterile surfaces onto which any transient microbes may attach and possibly grow if conditions are favorable[66,72]. Such growths may be examined visually or by various microscopic techniques. Such downhole recovery/incubation systems offer the advantage of replicating the conditions present in the wells water column but the disadvantage that the system has to be left downhole for the incubation period, which may extend into weeks or even months if an adequate time is to be allowed for attachment and growth. Following this, the unit has to be recovered (at the appropriate time) and observed.

Another approach to overcome the inconvenience of downhole incubation is to divert some of the water being pumped from the well through a device filled with porous media relevant to the conditions associated with the well. For example, a water well installed in a sandy aquifer would use a sterile sand medium. Water diverted through this porous medium would carry any transient microorganisms, some of which may attach or become entrapped in the porous medium. Two subsequent events would consist of physical (due to the entrapped particulates) and biological (due to the growth of biofilms within the porous medium) plugging[56]. Both would cause a loss in hydraulic transmissivity through the device and changes in the visible characteristics of the porous medium. The use of glass walls around the porous medium would allow a direct observation of these visible changes while changes (reductions) in the rate of discharge under standardized flow conditions could be used to record these effects[49].

Given that a water sample has been taken and there is a need to determine the potential for a microbiological presence to be occurring, a range of strategies can be followed. These are summarized below and subsequently the various relevant methodologies will be dealt with in detail.

Direct Inspection

A clear water sample has traditionally been considered to be one that is free from major microbiological problems (particularly when it is also coliform negative). Where there is an initial cloudiness which clears steadily from the bottom upwards, this may be due to dissolved gases (under pressure) becoming released (i.e., the cloudiness) as bubbles rise and disperse. Where a cloudiness gradually forms within the water sample after collection, this could be the result of microbial activity and growth in the sample itself. If the water sample came from an econiche where there had been an incomplete biodegradation of the organics because of such factors as anaerobic conditions, the taking of the sample would introduce aerobic conditions which would stimulate aerobic degradation. If the water is cloudy initially and it settles out leaving the water above clear, this may represent a denser type of particulate structure which may have originated from biofilms within the biofouled zone. Here, the density may be a reflection of the bioaccumulation of various organic and inorganic compounds together with the formation of "tighter" water-holding structures within the growths. If there is an accumulation of inorganic compounds, this might influence the color and texture of the settled material. Ferric (iron) and manganic salts will, where these are bioaccumulated, cause the settled material to vary from a yellow to orange to red to brown to black coloration. In addition, the settled material may become more "flake-like" when shaken up.

Another direct inspection involves the detection of any odors in the water. These smells are sometimes very characteristic for different microbial events. Common odors microbiologically generated include "rotten eggs", "kerosene-like", "earthy-musty" and "fishy". All of these when detected can indicate that particular groups of microorganisms have been active within those waters. Perhaps the most serious odor is "rotten eggs" which is associated with hydrogen sulfide production, anaerobic conditions and corrosion of equipment and installations.

Cloudiness is, of itself, an indicator of possible microbiologic problems. Standard approaches to evaluate this involves the measurement of the degree of interference or reflectivity that occurs when a beam of light is passed through the water sample. Decreases in the light intensity passing through the water (absorbance) or increases in the light being reflected as it passes through the water (reflectance) can be used to estimate the density of particulates in the water. Newer technologies utilize interferences that occur in a pulsed laser light passed through the water due to the presence of particulates. Laser particle sizing systems are not only able to count individual particles but also ascribe a size and even a shape. The typical range for such laser-driven systems is from 0.5 microns (less than the size of a typical vegetative bacterial cell) to as large as 100 microns or more. The total volume for the particles can be calculated and given as ppm (v/v) of total suspended solids (TSS). Often in cases where there is a biofouling event occurring, clustered particulate sizes may be recorded. These may typically be over a relatively narrow range of particle sizes such as (microns): 8 to 12, 10 to 15, 12 to 16, 14 to 18, 16 to 24 and 18 to 26. Where there are ribbon-like particles (e.g., stalks of *Gallionella*) present in the water, these will often cause spurious clusters of particles in the larger sizes such as (microns): 31 to 34 and 59 to 64. The laser-driven particle sizing offers the potential to "view" any suspended material down to the size of a bacterial cell or smaller. A water sample showing a < 0.009 ppm of TSS and recording no particles larger than 0.5 microns may be considered to be a water sample with a very low probability for the presence of microbial entities.

Indirect Inspection

Other techniques of microbiological examination involve a more indirect approach which uses either a generalized or differential staining, determination of metabolic activities related to active microorganisms, or enumeration of cells able to grow on selective culture media under specific environmental conditions.

In the last century there has been a movement through

the use of staining techniques to microscopically confirm a microbial presence in water, through selective cultural techniques to sophisticated "biomarker" systems which are able, under some circumstances, to detect the level of activity, viability or specific microorganisms of particular interest and concern (e.g., coliform bacteria) in water samples[10,12,43,62].

Historically, the focus of attention in groundwater microbiology has been influenced by the development of microbiological monitoring technologies. In the latter part of the nineteenth century, the occurrence of iron bacterial infestations in water wells and distribution systems triggered a level of effort to determine the causant agents and the factors controlling these occurrences. In the late nineteenth century, an applied microbial ecologist, Winogradsky, developed selective media and determined the cultural conditions which would support the growth of these organisms[18]. The elegant cell forms of some of the iron bacteria (e.g., the sheathed and stalked iron bacteria) led to these becoming a center of interest at that time. The ecology of groundwater systems and water wells did not develop significantly since the iron bacteria were considered to be nuisance bacteria of relatively little significance when compared to the very major significance of the coliform bacteria as hygiene risk indicator organisms.

Coliform detection methods were developed during the early part of this century and became well established so that now the presence of these bacteria is considered to be the prime indicator organism for hygiene risk in waters. This mindset established the notion that the coliform test was an indicator for all nuisance bacterial activities. Consequently the presence of coliform bacteria has often been taken to indicate that there is a bacterial problem while the absence of coliforms would indicate that there was not a microbial problem. Unfortunately, many of the bacteria associated with "nuisance" events in groundwater systems (e.g., plugging, corrosion, taste, odor and color problems) are usually not coliform bacteria. As a consequence of this, in practice, the absence of coliforms from a water has been commonly construed to indicate that there were no bacteria present. This would then be interpreted to mean that the problems observed were of a "chemical" nature rather than biological. Today it is well documented that microorganisms are prime factors initiating corrosion, plugging, and the accumulation or degradation of specific chemicals. Such microbiological events have traditionally been downplayed because of a well-meaning but, in this case, misdirected over-reliance on the coliform test as being an indicator of "nuisance" as well as "hygiene-risk" bacteria[50,51].

The net effect of this strategy of over-reliance on the coliform test has been that less attention has been paid to the ecological implications of microorganisms present and active in groundwater and groundwater-related engineered systems. A surge of research and developments was initiated in the mid-1970s and has progressed to date but with a low level of funded effort. In the 1980s, a surge of interest occurred generated, in part, by the growing recognition of the importance of managing the microbial components active within the hazardous waste control and biodegradation industry[6]. Additionally, studies have been conducted on deeper subsurface microbiology[28]. Present findings indicate that the crust of the earth is populated by microorganisms which have somewhat different characteristics to those found at or close to the surface of the planet. This "lays to rest" the second impediment to the recognition of microbial activity in groundwaters which is that "groundwater is essentially sterile". Here, the term "sterile" has been confused with the term "hygienically safe" (i.e., free from detectable coliform bacteria).

The net result of the recent initiatives and the overall

"rebirth" of microbial ecology is that the technologies for microbial monitoring and manipulation are in a state of growth. Traditional testing methods are rooted in developments that happened with the medical and food industries and were transferred as being applicable to the water industry.

The indirect technique most commonly used to enumerate bacterial numbers is the extinction dilution techniques followed by the spreadplating of the dilutions onto agar culture media in petri dishes. These are now incubated at an appropriate temperature and in conditions which would favor the targeted microbial group(s). Such "favored" organisms are expected to metabolize, grow and form visible distinctive clumps of growth called colonies. Each colony is thought to have arisen from a single colony-forming unit (cfu). Often the assumption is made that each colony arose from a cfu which was in fact one microbial cell. The net end product of this form of analysis is that each cfu counted is reported as the number of cfu which could be back calculated to have occurred in one milliliter or one hundred milliliters of water (i.e., cfu/mL or cfu/100 mL respectively). While there is comfort in obtaining a numerical value for the numbers of microorganisms of a specific type, there are some serious shortcomings in the application of this monitoring technique. These potential inadequacies relate to several aspects of the approach used and are discussed below.

Variability in cell numbers in a cfu

In waters, and particularly in groundwaters, much of the microbial cell population may be incorporated into particulate structures. The diameter of such particles may vary considerably from being as small as 4 to 8 microns, to as large as 80 to 120 microns. Small particles may contain either no viable entities, a single cell or a few cells dispersed or clumped together within the particle. Large particles could contain far more microbial cells since there would be a larger volume of occupancy. In the extinction dilution technique, these particles become diluted within the diluent (e.g., Ringer's solution). This dilution may cause shifts in the size of the particulate structures due to disruption (particles break up into a number of smaller units) or aggregation (particles lock together to form fewer larger units). Additionally, particles may become temporarily attached to the walls of the apparatus (e.g., pipette) leading to the loss of that particle from the water phase permanently (if the particle does not detach) or temporarily (if the particle does detach later). In the latter case, the particle could appear in a much greater dilution of the series to cause an anomalous cfu growth. All of these events may cause the spreadplate data to be artificially low or high[48]. While costly, the use of a fresh sterile pipette for each sequence in the dilution series would reduce the risk of this occurring.

Low cfu counts may be a reflection of the concentration of many of the viable entities within each particulate structure which may then produce only one colony incorporating the growth of one, some or all of the different types of microorganisms present in the particle. Additionally, low counts may also occur because some of the particulate structures have become attached to the surfaces of the diluting apparati and are no longer available to be recorded as cfu on the spreadplate surface.

Nutrient concentrations in the agar culture medium

The development of enumeration techniques was largely developed in the medical and food industries to determine the numbers of pathogenic and spoilage microorganisms respectively. These groups of organisms both infest environments which are generally very rich in nutrients, and consequently, the agar culture media were also high in nutrients. Commonly, the major sources of organic nutrients are added in the percentile concentration range of 0.1% to 5.0%. In groundwater systems, it is rare (except in the case of a severe organic pollution) to find environmental conditions in which the organics are in concentrations in excess of 0.001% (or 10 ppm). The exposure of such microorganisms from nutrient poor environments to the nutrient rich agar culture media can be traumatic or toxic. Such organisms are **organo-sensitive** (i.e., not tolerant to radically higher concentrations of organics). Here, negative responses cause the resultant enumerations to fall short of the true estimate of cfu. The trend in groundwater microbiology is to take into account these sensitivities and use agar culture media in which the nutrient concentrations have been reduced. Typically, the R2A agar medium used extensively in the enumeration of waters may contain only 10% or 1% (i.e., R2A/10 and R2A/100 respectively) of the nutrients normally recommended for use in this medium. Higher cfu recoveries can be achieved which reflect more accurately the number of viable microbial entities in the water sample.

Environmental Conditions for Incubation

While there is a well acknowledged selective effect through the type of agar culture medium used to enumerate microorganisms, there is another major consideration which relates to the conditions under which the charged agar spreadplates are incubated. The microorganisms in a groundwater sample have almost certainly come from an environment in which the physical (e.g., temperature, pH, redox) and chemical conditions are relatively stable. Additionally, the intense consortial nature of growth for these incumbents would heighten the interdependence of each component strain on the other members. Clearly, there would be a radical shift in the environmental conditions on the surface of the spreadplate. The shock that this may create can be lessened by minimizing the shifts that would occur during incubation. Some factors which are relatively easy to control include time, temperature, atmospheric composition and the composition of the culture medium. Time is perhaps one of the most perplexing factors because there is a natural conflict between the need to get the data as quickly as possible and allowing enough time for the incumbent microflora to adapt and grow. This period of time for adaptation can vary considerably depending upon the nature of the organism and the animation status (i.e., active or passive) of the cells. Animation status relates to the level of metabolic activity going on in the cells at the time of sampling. Cells that are actively growing and reproducing would be considered to be in a very active animation state while those which are totally passive would be considered to be in a suspended animation state. Generally, the less active (animated) the organism, the longer the time (adaptation phase) the organism will require before it begins to form a colony on a suitable agar culture medium. Since it is unlikely that the microflora will be in a very active adaptable animation state in groundwater, there is bound to be a significant adaptation phase before growth. The first cfu to appear on an agar spreadplate are likely to be those which had the shortest adaptation time. Other organisms would sequentially appear in order of increasing adaptation times. From the chronological perspective, the greater the delay (i.e., the length of the incubation period) before enumerating the spreadplates for cfu, the more complete will be the determination of the microflora.

Expediency often dictates a rapid "turn around" in the enumeration of microorganisms. This incorporates a "risk" that all of the incumbent organisms able to grow under the conditions of the test would not have had enough time to do so. In a radical condition where all of the microflora are in a low state of animation, a test may register 0 cfu/mL simply because the incubation period has been too short to allow these organisms to adapt, grow and develop into colonies which can then be enumerated. Prolonged incubation for spreadplates is advised beyond the periods commonly practised (e.g., 2, 5 or 7 days). Experiences in Regina would suggest that enumerations should extend to a minimum of 14 days with 21 or 28 days of incubation continuing to show a greater number of cfu present. Often the number of colonies rises in a step-like manner over periods as long as six weeks (42 days).

Temperature is another major physical factor which can affect the range and rate at which the microflora adapt and grow on the agar culture medium. A common feature of groundwater systems is that the ambient temperature does not fluctuate to any great extent on a daily or seasonal basis when compared to surface environments. The net effect of this is that the incumbent microflora has adapted to these narrow ranges of temperature. Enumeration involves a number of stages which can be, to a varying extent, traumatic to the incumbent organisms. Typically these stages would include temperature shifts during sampling, storage, preparation of the groundwater sample for testing and, finally, the temperature of incubation. Each of these "stages" could, by the nature of the temperature shift occurring, cause a traumatization of at least some component within the microflora.

For example, groundwater is sampled from a formation in a biozone operating at 8.3°C. This water mixes with groundwater from other formations during entry into the water well and subsequent pumping. This causes the temperature to elevate to 11.6°C which is the temperature of water sample when taken. The ambient air temperature is 24.6°C and the water sample temperature rises to 14.1°C before the water is stored, packed in ice, in a cooler. Here the temperature declines to 7.5°C over six hours. Once in the laboratory, the water sample is left standing for two hours at room temperature (22°C) before the extinction dilution is carried out. Water temperature rises to 17.4°C. The technician now performs the dilutions using sterile Ringer's solutions just removed from the refrigerator (4°C) which cools the water down to 8.3°C. After five minutes the diluents are spread out over the selected agar culture medium which had just been removed from the refrigerator (4°C). The cell temperature falls rapidly to 4.2°C. Once the plates were moved to the incubator (30.3°C), the cell temperature rose rapidly to that of the incubator. After five days, the colonies were counted. This example shows that the microorganisms would have been subjected to a series of temperature shifts which would not have been experienced in the natural setting. Not surprisingly there would be a reduced potential for such organisms to grow under the conditions presented on the surface of the spreadplate.

In essence, the most appropriate manner to handle a sample, where there is a relatively short storage and transit period, would be to keep the sample at as close to its original site temperature as possible. In the case of the example above this would be 8.3°C. Secondarily, the temperature selected for the enumeration should be within the same general range as that of the original habitat. In the example shown above, 30°C was used to incubate the sample. This temperature is outside the immediate range of temperatures around which an organism naturally functioning at 8.3°C may always be expected to function.

As a guideline to the selection of suitable temperatures for the enumeration of the natural microflora functioning at a given site, the following structure can be employed (temperature range of the habitat,°C; incubation temperature range, °C low - high): (2 to 10°C; incubate at 8 - 25°C); (11 to 15°C; incubate at 12 - 25°C); (16 to 30°C; incubate at 25 - 30°C); (31 to 40°C; incubate at 35 - 37°C) and (greater than 40°C; incubate within 5°C of the natural source temperature).

It is generally thought that the higher the selected incubation temperature, the faster the rate of growth. However, in groundwater situations, given the alien nature of the enumeration procedure, it would be preferable to allow a minimum of seven days incubation (longer if possible).

Where there is an urgency to obtain data, early colony counts may be performed and the data referred to as **preliminary cfu/mL**. Such data should be "backed up" with **confirmatory cfu/mL** calculated from the number of colonies obtained after prolonged incubation where the colony numbers would, by repeated counting, be found to have stabilized. In general, a "plateau" in colony numbers may be reached in 21 to 42 days.

pH is a physical parameter that is well known to influence the types of microorganisms which may be grown on a spreadplate[78]. Most microbial activity commonly occurs within the range from 6.5 to 8.75 pH units with optimal growth frequently evident at pH values from 8.2 to 8.4. Most culture media tend to have the pH adjusted to within this range with 7.2 to 7.4 pH units commonly being employed. Where the pH of groundwaters falls outside (commonly more acidic) the normal range for microbial activity, organisms within a biofilm often do have the ability to "buffer" the pH shift[58]. This allows the pH of the biofilm itself to remain in a suitable range for the activities of the consortium of organisms present within the film. When groundwater pH values are acidic but not below 3.5, there is a possibility that a "buffered" biofouling may still be occurring in association with that system. Regular culture media may therefore be adequate to enumerate these organisms. For the more acidic groundwaters (pH < 3.5), there is likely to be a dominance of acidotrophic bacteria growing and specialized media will be required to enumerate these organisms.

Microorganisms are often very sensitive to the reduction oxidation (redox) potential of the environment in which they are growing. For many of the groundwater microorganisms, conditions are borderline between oxidative and reductive with oxygen in short supply (often ranging from 0.02 to 2.5 mgO_2/L) and respiration switching between oxygen and nitrate as the terminal electron acceptor or turning off if there is no substrate. In the typical aerobic spreadplate count, the surface of the agar is now exposed to atmospheric oxygen which saturates the surface agar strata. These conditions are more oxidative than the microorganisms may have been accustomed to growing in. As such there may be a period of adjustment prior to organisms growing in this very different environment. The atmosphere around the spreadplate can be modified prior to incubation to render a reductive condition through the removal of oxygen from the incubator. On occasions, the replacement gas may be a combination of carbon dioxide, methane, nitrogen and hydrogen in various ratios. These conditions now allow many of the anaerobic oxygen sensitive microorganisms to grow and form countable colonies. The selection of an "open" incubation will encourage aerobic growth and exclude the growth of any oxygen sensitive organisms (i.e., the non-aerotolerant anaerobes). Selection of a replacement reductive atmosphere will shift the growth forms to some fraction of the anaerobic population. The nature of these growths would be dependent upon the nature of the gases and culture medium used.

Traditionally, the agar base used in the "solid" medium has been accepted because of the convenience created by: (1) the distinctive colonial growths which evidence the visible presence of each cfu; and (2) the agar is not commonly degraded by microorganisms. This latter factor has supported the widespread adoption of agar for the evaluation of the microbial components in environmental samples. Only rarely will the agar be found to "liquefy" because of the activities of these "agarolytic" microorganisms. The convenience and historic use of agar-based culture media has led to the acceptance of data gathered by this route without perhaps an adequate consideration of the various limitations, some of which are discussed above[46]. Frequently, the presence or absence of microbial activities is based on the application of a spreadplate evaluation rather than on a broader spectrum evaluation of microbial activities by a variety of direct and indirect techniques. The next sections of this chapter address the methodologies which can be employed to monitor the potential for microbial occurrences.

CONCEPTUAL APPROACH TO MICROBIAL EVALUATIONS

There are two fundamental methodologies for the evaluation of a microbial presence within a groundwater system, direct and indirect. Direct examination involves the determination of biomass presence at the molecular, cellular or conglomerate (particulate and biofilm) level[69,76,79]. Indirect examination involves the determination of viable entity presences through subsequent cultural and/or metabolic activities of various types.

Traditionally, the concept has involved enumeration of the viable entities present within a given water sample expressed as either **viable units (vu)** or **colony-forming units (cfu)**. The volumes commonly used for these determinations are either one milliliter **(/mL)** or one hundred milliliters **(/100mL)**. Where the concern is hygiene risk (potential presence of pathogenic microorganisms), the more sensitive scale (/100 mL) is commonly employed. For the "normal" microflora within the environment, there is a greater tolerance for these populations and these are normally expressed per milliliter (/mL). The term cfu is normally attributed to any enumeration involving the enumeration of colonial growths on an agar culture medium. This may be a generalized enumeration in which all of visible (by eye or low magnification stereo microscopy) may be counted **(total colony count)**. Alternatively, a selective enumeration may be undertaken where only the **typical colonies** are included based on the specific criteria documented as appropriate to the selection procedure. Other colony forms that do not follow the criteria established for the typical colony count may be enumerated separately as **atypical colonies.**

Viable units (vu) do not involve the formation of distinctive colonies but rather the generation of some product of growth (e.g., cells to develop turbidity; organic acids to lower the pH). By sequentially diluting the water sample (usually in tenfold series or a statistically appropriate manner) in sterile culture media which will allow the required product to be generated where viable units were present. By noting which dilutions reacted because of the presence of active viable units, it becomes possible to calculate statistically the theoretical number of vu/mL or vu/100 mL. The most common application is in the determination of total coliform and fecal coliform bacteria by the **most probable number (MPN)** method.

Aggressivity is another method for determining in a semi-quantitative manner the population size and activity level of a targeted microbial population. In this method, the water sample is placed in an environment (e.g., liquid selective culture medium) under incubation conditions which would encourage the growth and activity of the targeted microorganisms (e.g., iron related bacteria). Here, the rate at which the onset of a determinable event (e.g., gas production, color generation, clouding) occurs is considered to relate to the aggressivity of the organisms within the sample. For example, a very small population of microorganisms which are very active and rapidly generate a determinable event would be more aggressive than a large population of microorganisms which are in a passive state and do not respond to the test environment so quickly. It may be argued that, in this case, the small population has much more aggressive potential than the passive larger population and, as such, presents a greater fouling or hygienic risk.

SIMPLE FIELD TEST METHODS

When working in the field remote from laboratory support facilities, there has developed a range of simple techniques which can give an indication of whether there is a microbially driven event occurring (e.g., MIF, MIC). These techniques are by nature very simple and in some cases involve locally available material. Incubation is at room temperature as a matter of convenience.

Rodina Test

This test was described in 1965 and it functions through the ability of some IRB to generate a flocculant type of growth when the water is subjected to very aerobic conditions[66]. The test (Figure 33) itself consists of pouring water into a wide-necked flask or bottle and leaving the water to sit overnight. Do not fill the container to a depth of more than 10 cm. Here, the wide neck of the vessel generates a large surface area through which oxygen can diffuse into the water. This increasing oxygen presence together with a non-turbid environment causes some IRB to form into flakes which resemble discolored cotton wool in appearance. Where this occurs, the presence of IRB can be suspected and confirmed conveniently by a direct microscopic examination of the "growths".

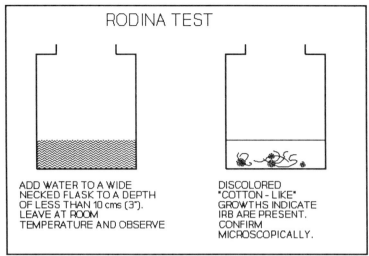

Figure 33. *Rodina[66] developed a very simple test in which a shallow layer of water (left) was left at room temperature. The aerobic conditions triggered IRB development (left).*

Cholodny Test

In 1953 this method was described in which the IRB could be seen if these organisms attached to glass (Figure 34). The method recommended that some of the water and any sediment recovered be placed in a jar. A cork is then floated on the surface of the water with several cover glasses (slides or cover slips could be used as alternatives) attached on the underside. As the water clears through sedimentation, the glass attached to the cork becomes visible. The development of any "rust spots" or "cotton-like accumulates" either on the surface of the sediment or on the glass inserts is taken to indicate the presence of IRB. Where there were very large and aggressive populations of IRB, such events could occur in less than 24 hours of incubation. Where identification of the IRB was desirable, the glass inserts could be removed with the cork from the water, detached and air dried. Upon staining using one of the recommended

IRB procedures, the types of IRB present and active in the sediment - water can be identified.

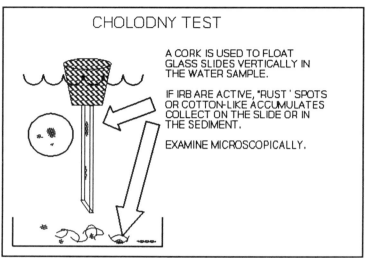

CHOLODNY TEST

A CORK IS USED TO FLOAT GLASS SLIDES VERTICALLY IN THE WATER SAMPLE.

IF IRB ARE ACTIVE, "RUST" SPOTS OR COTTON-LIKE ACCUMULATES COLLECT ON THE SLIDE OR IN THE SEDIMENT.

EXAMINE MICROSCOPICALLY.

Figure 34. *Another simple test (Cholodny[60, 66]) floated a glass slide (left) in the water sample by a cork. IRB presence would occur through visible growth either on the slide or below.*

Grainge and Lund Test

This test identified with the need to develop a very simple test in which iron is added to the water sample in such a way that the activity of IRB could be easily identified[37]. This technique was described in 1969 and recommended as a monitoring procedure for the effectiveness of control programs. A clean soft steel washer (preferably chemically cleaned) is placed in a conical flask (e.g., Erlenmeyer flask) along with an extruded plastic rod (e.g., stir stick) which is positioned so that is vertical (Figure 35). Water is added sufficiently to cover the washer and leave the end of the plastic rod sticking out of the water. After being left for two days, the water line around the plastic rod is examined for any translucent string-like (filamentous) growths. Over time this growth develops a brown tinge (iron accumulation). This is taken to be a positive indication of the presence of IRB in the sample.

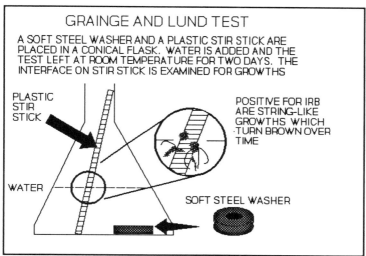

Figure 35. *In the Grainge and Lund test[37], a plastic stir stick forms the growth support structure for the IRB. Growth may be seen on the stick at the water-air interface.*

GAQC Test

Otherwise known as the George Alford Quick and Crude test, this test involves the use of materials that are easily at hand to determine whether IRB are present or not. The method, developed in 1980, involves the admission of a non-galvanized iron washer to the water sample along with a carbon source (ethanol) to stimulate the activity of the IRB. To test a water sample, 150 mL is added to a clean glass and the washer dropped into the water to come to rest at the bottom. Two drops of Jack Daniels whiskey are added to the water ("one drop for me and one drop for the bugs", George Alford, personal communication, Figure 36). Cover the glass loosely with aluminium foil (to reduce evaporation) and incubate on top of a refrigerator (warm) or on a shelf or windowsill. After two days a positive reaction is recognized by either a "fuzzy" growth around the washer or metallic "floaters" in the water. In both cases IRB are actively either attaching directly to the source of iron (i.e., the washer) or accumulating the dissolved iron within suspended particulate masses (i.e., the floaters).

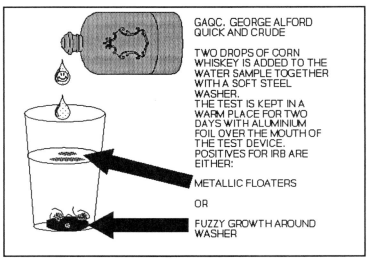

GAQC. GEORGE ALFORD
QUICK AND CRUDE

TWO DROPS OF CORN WHISKEY IS ADDED TO THE WATER SAMPLE TOGETHER WITH A SOFT STEEL WASHER.
THE TEST IS KEPT IN A WARM PLACE FOR TWO DAYS WITH ALUMINIUM FOIL OVER THE MOUTH OF THE TEST DEVICE.
POSITIVES FOR IRB ARE EITHER:

METALLIC FLOATERS

OR

FUZZY GROWTH AROUND WASHER

Figure 36. *A more recently devised simple test utilizes whiskey (as a carbon source), an iron washer (source of iron) to generate IRB growth either floating on the surface, or growing in the base.*

BART™ Test (Waverley Scenario)

Since 1988, the Town of Waverley in Tennessee has been using the BART™ biodetectors to determine in a simple manner whether there was a recurring MIF problem developing in the water wells[1]. In the simple scenario developed at this time, tests were performed monthly on the water from the wells using the W series biodetectors. Experience found, in this case, that it was the three units (SLYM-, IRB-, and

SRB-BART™) which could be used to monitor a potential biofouling problem (increased drawdown and unacceptable increases in manganese concentrations). The key factor was the days of delay to a reaction being generated. Where the system was functioning efficiently, the normal *dd* was at between 12 to 14 days. When the *dd* fell below 12 days there was, through practical experience, an increased occurrence of biofouling. Normally it would be the SLYM- and the IRB- which would drop first to 9 or 10 days. To check this, additional water samples were drawn on two consecutive days and tested. If these both showed the same foreshortened *dd*, there would be an increased risk of biofouling and corrective action (e.g., shock chlorination or acidization) would be applied. Tests were repeated after treatment. A successful treatment would be deemed one that had caused the *dd* to return to the normal level (i.e., > 12 days). It was found that where the SRB- BART™ also showed a shorter *dd*, the treatment had to be more intensive to return the wells back to an acceptable *dd*. On some occasions the increase in aggressivity of the SRB (i.e., shorter *dd*) preceded rapid increases in the manganese concentrations in the water. This may have been due to a greater amount of sloughing from a more matured biofouling in which the more deeply seated SRB components were being released along with more of the accumulated manganese.

The Waverley scenario offers a very simple comparative way of monitoring the status of biofouling within a water system using the premise that if the *dd* is getting longer (everything is getting better), if the *dd* is constant (everything appears to be stable) and if the *dd* is getting shorter then the microorganisms are becoming more aggressive. In the latter case, duplicate samples on two consecutive days should be undertaken to confirm that there is a decline in *dd* (i.e., the original test was not a "chance" event). Once confirmed by the duplicate tests, treatments should be undertaken to remediate the MIF occurrence. It should be noted that the patented BART™ biodetector system was developed by Cullimore and Alford from the MOB (metallo-oxidizing bacteria) and the MPNB (metallo-precipitating non-oxidizing bacteria) test system described by Cullimore and McCann in 1978[18]. At that time the test method had been designed for simple field use and differentiated six reaction types. Today, that test method is incorporated into the IRB-BART™ biodetector. It was found during that research phase that 95% of the groundwaters tested in Saskatchewan, Canada were positive for IRB. Microscopic examination found that, of the sheathed and stalked IRB, *Crenothrix, Leptothrix, Sphaerotilus* and *Gallionella* were frequently dominant types. Given the universal nature of the presence of IRB it becomes more critical to appreciate their relative aggressivity which can be measured by the *dd* characteristic.

MICROSCOPIC INVESTIGATION FOR IRON RELATED BACTERIA

Olanczuk-Neyman Method

This methodology was developed in Poland while working on IRB fouling of water intakes. A staining technique was employed which used membrane filtration to concentrate the suspended particulates prior to staining with a modification of the Rodina procedure.

1. Separately dilute 10, 1 and 0.1 mL of the water sample into 110, 119 and 199.9 mL of sterile Ringers solution respectively to create final volumes of 120 mL. Gently mix by rotating each dilution in turn.
2. Filter each diluent through a separate 0.45 micron membrane

filter and fixate the IRB to the filter by immersing each filter into a 1.0% solution of formalin for a minimum of 20 minutes.
3. Following fixation, immerse the filter in a 2.0% solution of potassium ferricyanide for 20 minutes.
4. Rinse the filter off in 5.0% hydrochloric acid for 3 minutes.
5. Wash the acid off the filter gently using distilled water.
6. Cover the filter with a 5.0% solution of Erythrosin A in a 5.0% phenol-based solution. (Note that a 2.0% solution of Safranin can be substituted for the Erythrosin A)
7. Gently rinse the membrane filter with distilled water to wash off the stain and air dry.
8. Examine microscopically.

Under low power magnification, the **bacteria cells will appear to be stained red** while the **iron deposits will stain blue.** Under oil immersion examination, the membrane filter will become transparent when saturated with the oil. This makes the observation of the IRB much easier. The Olanczuk-Neyman method calls for the numbers of IRB to be enumerated using 100 microscopic fields selected randomly and this is related back to the number of IRB/mL using the formula given below. It should be remembered that much of the material being viewed microscopically is dead, including the stalks of *Gallionella*. Care must be taken to determine the location of viable cells.

To calculate the IRB/mL, the following formula can be used:

Diameter of the microscopes field of view: A microns
Average number of IRB on 0.1 mL dilution fields: B
Average number of IRB on 1.0 mL dilution fields: C
Average number of IRB on 10 mL dilution fields: D
Number of fields counted at 0.1 mL dilution: E
Number of fields counted at 1.0 mL dilution: F
Number of fields counted at 10 mL dilution: G

Counts are recorded where IRB cells were present, fields observed which did not display any IRB are also included.

Calculate the IRB/mL for each dilution field:
    H = B * 10
    I = C
    J = D * 0.1

Total number of field counted:
    K = E + F + G

Since the fields reflect only a factorial part of the total surface area, there needs to be a correction factor which would allow the total population of IRB entrapped on the filter to be calculated (and representing the total volume of water sample used). This factorial (L) may be calculated as the relationship between the area (AV) of the field of view (diameter A given) and the area (AM) of the membrane filter (diameter M given):

    AV   = $3.142 * (A/2 * A/2)\ 10^{-12}\ m^2$
    AM   = $3.142 * (M/2 * M/2)\ 10^{-12}\ m^2$
    L   =   AM / AV   factorial area increase.

Averaged Population of IRB (PIRB):
    PIRB = (H * (E/K)) + (I * (F/K)) + (J * (G/K)).
The **APIRB** is expressed as IRB/mL.
    APIRB =   PIRB * L.

Meyers Stain (modified by Pipe)

Meyers described a staining technique for IRB in 1958 which has been found in the experience of Pipe to be superior to the other techniques presently available[18]. Like the Olanczuk-Neyman method, this technique stains **the bacterial cells red** while the **iron deposits are blue**. This technique is primarily recommended for use with water

samples which are showing some signs of a possible infestation (e.g., the water is discolored to a yellow, orange or
brown hue, the water has a distinct colored deposit which
may have an indistinct or fuzzy outline).

1. Separate the cellular material by centrifugation to the extent
that a pellet is formed in the base of the centrifuge tube. The
precise level of centrifugation will vary since a greater amount
of g force (gravity) may need to be applied to "draw down"
the cells in some waters.
2. Pipette off the water carefully so as not to disturb the
pellet.
3. Using a sterilized clean bacteriological loop, withdraw
some of the pellet and smear onto a bacteriological slide.
4. Air-dry the slide and heat-fix by passing the slide quickly
three times through the lower part of a blue bunsen flame. Be
careful not to overheat the slide. Only enough heat is required
to make it feel warm to the touch and cause the bacteria to
become "fixed" to the glass.
5. Immerse the slide in absolute methanol for 15 minutes.
6. While the slide is immersed, bring to boiling point a 1 : 1
mixture (v/v) of 2% potassium ferricyanide and 5% acetic
acid both as aqueous solutions in distilled water.
7. Remove the slide from the methanol and immerse in the
boiling mixture for 2 minutes.
8. Remove the slide and allow it to cool.
9. Gently wash the slide with distilled water.
10. Cover the slide with an aqueous 2% safranin solution and
leave for 10 minutes.
11. Rinse off the stain with water, air dry and examine microscopically.

This technique is not quantitative but does allow a
variety of the IRB to be identified as present (or absent) and
a relative relationship to be expressed between the species
observed. The limitation of the technique is that it does not
necessarily allow a clear observation of the CHI bacteria

which may have accumulated iron within irregular amorphous structures. Additionally the technique is not suitable for waters with a low IRB population. Such waters are best stained using the Olanczuk-Neyman or the Leuschow and Mackenthum membrane filter methods.

Leuschow and Mackenthum Direct MF Technique

This technique was described in 1962 for the investigation of waters which had a low number of IRB[57]. Unlike staining techniques, this technique involves the direct observation of the bacteria that have been entrapped on the membrane filter which has been dried and rendered transparent with immersion oil. It is restricted, as a technique, to only those bacteria which are occurring in large and/or distinctive structures and are pigmented (commonly by the entrapped iron oxides).

1. Filter 100 mL of the water sample through a 0.45 micron membrane filter.
2. Remove the filter and dry by exposing the filter to 100°C long enough for the filter to appear to be "bone dry".
3. Saturate the dried membrane filter with immersion oil. Since the oil has the same refractive index as the filter, the disc filter will become transparent.
4. Place the filter on a bacteriological glass slide and examine under high power (x400 to x600 diameters).

Various common forms of stalked and sheathed IRB may be easily seen by this technique even when they are present in the water in relatively low numbers. In Wisconsin, Leuschow and Mackenthum recovered IRB from 55% of the well waters tested with *Gallionella* and/or *Leptothrix* being the most common dominant IRB. In the reddish turbid waters, counts by this method reached $> 10^7$ IRB /mL.

Negative Wet Mount Stain

One major problem with the staining techniques is that the "edge" of bacterial cells, particles, sheaths and stalks may be diffuse and difficult to observe. Additionally, where the bacteria are not pigmented or have not accumulated iron, the cells may be transparent to light and so not be recognizable. The negative stain uses a different concept in that the background is stained while the bacterial cells remain unstained[43]. Thus the cells stand out as being illuminated zones within a darkened (stained) background. This is achieved using **nigrosin** which is an acidic stain. Such stains do not penetrate and stain bacterial cells due to both the stain and the bacterial cells being negatively charged. The stains do tend to produce a deposit around the cells which forms into a dark background in which the bacteria appear as clear, unstained regions which relate to the shape of the cells. Where nigrosin is not available, India ink (25% aqueous solution) may be used.

This negative stain is more satisfactory for waters which are showing some visible signals of possible microbial presences (discoloration and clouding) and has the advantage that all types of bacteria may be observable by this technique. There are a number of mechanisms to obtain the sample for the negative stain. Approximately 0.2 mL of suspension is used in this staining technique. A very turbid water or a centrifuged pellet would probably have too many cells to be clearly definable. A regular "clear" water sample may have too few a number of cells to be conveniently observed. In either event, some dilution (into a sterile Ringers solution) or concentration (by passive settlement or centrifugation) may be needed to conveniently view a dispersed microbial mass. Trial and error may be required to find the appropriate dilution or concentration of the water sample.

1. Using a sterile inoculating loop, transfer two drops of the sample to the surface of the clean glass slide.
2. Add two drops of nigrosin and evenly mix the sample with the nigrosin.
3. Spread the mixture evenly over the surface of the slide using the narrow flat edge of a second slide held at an angle to the first slide. As the edge is moved along over the surface of the glass slide, so the suspension should spread out in its wake to form a narrow film.
4. Lower a cover slip of the central area of the film and gently lower one end first using forceps. Gradually reduce the angle of the cover slip until it is totally in contact with the film. Take care that no air bubbles are entrapped under the slide.
5. Using high power and oil immersion assisted microscopy, examine the film through the cover slip.

Cells will appear to be either transparent (if there has been no excessive accumulation of iron) or orange to brown (where iron has been accumulated). The background will appear to be pale blue. Where there is a "tight" form of ECPS around the cell (e.g., capsule, sheath or a thick coat of ECPS), this will form the distinct boundary rather than the cell itself. With careful focusing and adjustment of the light, the cells can sometimes be seen within the extracellular structures.

A **dry mount** negative stain can be made as a "library" copy which can be kept and viewed at leisure (the wet mount preparations will dry out). Where this is desirable, the slide should be left to air dry after step 3 above.

Gram Stain

The Gram stain (gRAM) is a standard staining technique used for the preliminary identification of bacteria based upon **R - reaction, A - arrangement** and **M - morphology**. These characteristics relate to the staining reaction that occurs (R), the form in which the cells are arranged (A) and the shape of the cells (M). Generally, this staining technique is used when working with pure cultures or natural samples where there is a high density of cells and relatively little interference factors (e.g., ECPS or inorganic deposits). It is considered to be a **differential** stain because there are two major reactions: (**pink-red - gRAM negative, G-; blue-violet - gRAM positive, G+**)[62]. Waters commonly have either too dilute a bacterial community or too many interference factors to allow the direct application of the gRAM stain to a water smear. This stain method, however, is commonly used in confirmatory procedures where microorganisms have been grown in liquid (**broth**) cultures or on solid (**agar**) media as colonies.

There are many variations for the gRAM staining technique. The one described below is the technique recommended by Harley and Prescott in 1990[43].

1. Prepare a smear of the colony (using inoculating needle and dispersing the cells in a drop of water) or suspension of a broth culture over a glass slide. Heat fix (pass slide quickly through a bunsen flame three times), cool and label slide.
2. Flood the slide for 30 seconds with crystal violet and then rinse off the stain gently with water for 5 seconds.
3. Now cover the smeared part of the slide with Gram's iodine mordant and allow it to stand for 1 minute. Rinse off gently with water for 5 seconds.
4. Gradually (drop by drop) cover the smear with 95% ethanol. Note that violet streaks will begin to form as the crystal violet - iodine stain complexes bleed from the smear. Continue to apply the ethanol until the bleeding stops (if in less than 30 seconds). Rinse gently with water to remove the ethanol. Prolonged exposure to the ethanol may cause too much decolorization and even the G+ cells might then become affected to appear as G- cells.
5. Apply gently the safranin stain until it covers the smear and

leave for 1 minute. Rinse gently with water, blot dry with an absorbent (bibulous) paper.
6. Examine under oil immersion (x 1,000 magnification).

If there are too many cells to clearly see each individual cell then repeat the procedure but use less cell inoculum and /or a more diluent. If too few cells, scan the stained smear microscopically to investigate whether there has been any clumping of the cells. If not, repeat the procedure but use a greater amount of colony or do not spread the water sample across the slide and allow longer for the smear to dry before being heat fixed. It should be noted that many of the IRB do not gRAM stain well because of the interference factors which prevent the stain and decolorizing ethanol to penetrate to the cells. Additionally, yeast and molds, where these are present, routinely gRAM stain as G+.

SPREADPLATE ENUMERATION

"Microbial ecologists are often faced with the serious problem of large discrepancies between microscopic and plate counts, especially for cells isolated from natural environments. Using a sample from either soil or aquatic environments, the former count is usually 10 to 1,000 times larger than the latter count." Tsutomu Hattori[46].

The most common technique used and generally recognized as being valid is the extinction dilution/spreadplate enumeration (as cfu/mL or /100 mL) using a variety of agar media. There has, over the past two decades, been a growing debate as to which of the vast array of culture media are the most suitable. This debate has centered on the most appropriate culture medium for the enumeration of aerobic heterotrophs, hygiene risk indicator bacteria (e.g., coliforms, enterococci, *Clostridium welchii* and species of

Klebsiella, Acinetobacter and *Aeromonas,* iron bacteria, biodegraders, sulfate reducing bacteria and anaerobic bacteria). Considerations also range through the use of the MF technique (where there are low populations in the cfu/100 mL range) and the correct methodology for extinction dilution (where there are high populations in the cfu/mL range).

Methodologies[43] have tended to ignore the serious concerns expressed above with respect to the discrepancies observed between microscopic and spreadplate counts. These concerns would relate to the potential radical underestimation of the microbial cell populations in a water sample using the spreadplate techniques. However, a heavy reliance has been placed upon the relative value of the population numbers recorded in a sequence of samples and the cfu/mL generated has been considered by many users to closely reflect the number of cells.

In conducting a spreadplate enumeration from a water sample, there are a number of stages that are involved after the selection of the appropriate agar culture medium: (1) vortexing to disperse particles (optional); (2) serial dilution of the water; (3) dispensing of diluents onto the agar surfaces; (4) drying of the agar surface (optional); (5) incubation under the correct temperature and atmosphere (reductive or oxidative); (6) counting the colonies which are visible; and (7) computing the cfu/mL or cfu/100mL. Each of these stages will be addressed in turn.

Selection of the Appropriate Agar Medium

Agar-agar itself is a galactoside obtained from certain marine red algae (seaweeds). Most microorganisms are incapable of degrading agar-agar. It is used at concentrations of between 1.0 and 1.5% (w/v) depending upon the formulation of the medium and the quality of the agar-agar used. Lower quality agar-agar contains up to 0.5% phosphorus (as P w/w) while higher qualities contain less than 0.2% P. This translates into a potential supplementation of the agar medium with up to 75 and 30 mgP/L respectively. This would form a significant additional nutritional base for microbial growth and activities. Indeed, water agar utilizes the nutrients inherently present in the agar-agar and the water used to make the medium. On occasions where the microflora has a significant (if not dominant) organo-sensitive component, the water agar may give the highest counts of cfu/mL when compared to other recommended media.

Agar-agar was originally selected because of two very favorable factors. Firstly, the agar-agar was not degraded by many types of microorganisms and therefore relatively stable (when compared to gelatin). Secondly, the agar-agar would remain as a "solid" gel at temperatures up to boiling point (i.e., >90°C) which would allow incubation in the thermophilic range; and yet would not set from the molten state until 46°C. This latter feature allows microbial suspensions to be mixed into the setting agar-agar so that a greater dispersion occurs in the colonies that subsequently form. This is known as the **pour plate technique**. This method is not widely used for environmental samples because the heat shock caused by adding the organisms to cooling molten agar-agar can be traumatic.

In the cultural techniques using agar-agar, the water and supplemented chemicals are dispersed. For a colony to form, the organisms have to "mine" both the water and these chemicals from the agar-agar base. If organisms are not able to extract adequate water and nutrients from the gel base, colonial growth would not occur. This may be a major factor restricting the range of microorganisms able to grow on agar culture media.

In relative humidities where the agar medium is not subject to radical evaporation rates and drying out, 15 mL is a common volume to dispense an molten agar medium into a standard petri dish. However, in dry climates such as is commonly experienced in the American prairies, a greater volume of agar (20 to 25 mL) may be dispensed to compensate for the greater losses. "Bagging" the poured plates in a plastic bag may not function well since water may condense on the walls on drip back into the dish causing contamination and erroneous results. Inoculated agar plates are incubated upside down (agar facing downwards) so that water cannot condense on the upper surfaces and drip back down onto the growing colonies.

Each agar culture medium is developed to enumerate a particular spectrum of the microflora in the water and no single medium is capable of stimulating the active growth of all of the microorganisms that could conceivable be present. There are therefore broad spectrum and selective forms of agar media which serve a very different purpose. The broad spectrum media generate a set of nutritional conditions in which a wide variety of microorganisms may be able to flourish (e.g., the heterotrophic bacteria). Selective culture media implement the use of inhibitory chemicals (e.g., antibiotics), marker dyes and restrictive nutritional regimes which encourage and support the growth of a narrow spectrum of microorganisms. The various agar media are listed below.

AA agar: A very selective medium for the gRAM negative strictly aerobic heterotrophic bacteria (section 4) and, in particular, the genus *Pseudomonas*. Colonies are commonly discrete and may be pigmented or if the pigments are water soluble these may diffuse into the underlying agar medium.

Brain heart infusion agar/4 (BHI/4): This quarter strength brain heart infusion agar has been found to support excellent colonial growths of a broader spectrum of bacteria than the R2A agar medium[22,51]. Colonies tend to grow more rapidly and to a larger size. It is a very useful medium to obtain a generalized enumeration of aerobic bacteria. If placed under anaerobic conditions (e.g., through the use of an anaerobic jar and gas pack), colony counts for anaerobic bacteria can be obtained.

Czapek-Dox agar (CD): This is a broad spectrum medium for culturing many of the **fungi (molds)**. This medium is particularly suitable for the culture and enumeration of *Aspergillus, Penicillium* and related fungi. Heavy fungal populations may be present in groundwaters which have interfaced with unsaturated zones (e.g., via recharge) with some level of organic input (e.g., from sewage lagoons or oxidation ponds). Typical fungal colonies appear to have a rough, almost cotton-like appearance which spreads out over the surface of the agar. Visibly distinctive bodies within the mass of growth (called a mycelium) may form the sporing bodies.

LES Endo agar: This medium has been developed for the enumeration of **total coliforms (TC)**. It is a complex medium containing sodium desoxycholate (a bile salt) which will inhibit many of the bacteria which are not commonly found in the intestinal tract. Lactose is the major carbohydrate (sugar) which is fermented by a relatively narrow spectrum of bacteria which includes the coliforms. Typical coliforms have a colony with a golden-green metallic sheen which develops within 24 hours. This sheen may be centrally located, around the periphery or cover the whole colony. It is often used as a basal agar for the incubation of membrane filters. It is generally considered that a valid colony count will be one where there are between 20 and 80 typical colonies per filter and no aberrant atypical overgrowths.

Lipovitellin-salt-mannitol agar, is a medium used to determine the presence of *Staphylococcus aureus*. This bacterium is commonly a concern in confined recreational waters where bathing is practised. When present, the colonies formed are surrounded by a yellow zone with opacity. The opaqueness relates to activity correlatable to the presence of coagulase-positive staphylococci.

M-FCIC agar: This is a medium which is specific for the genus *Klebsiella*. Typical colonies of *Klebsiella* produce a blue to blueish grey colony. This medium is used to determine the presence of this genus in waters by membrane filtration. Atypical colonies are a beige to brown color. *Serratia marcescens* will grow but will be distinctive through producing a pink colony.

M-HPC agar: This is an agar medium which has been developed for use with the **membrane filtration evaluation for heterotrophic bacteria**. It is an unusual medium in that it contains glycerol and gelatin as two major sources of nutrients.

Mn agar: This medium incorporates both the ferrous and manganous forms of iron along with citrate to encourage the growth of **iron related bacteria**.

M-PA agar: This is a medium used in conjunction with membrane filtration to detect the presence and numbers of *Pseudomonas aeruginosa*. Typical colonies are flat in profile with brownish to greenish-black centers which fade out towards the edge of the colony.

Pfizer selective enterococcus (PSE) agar: This is a medium which allows the differential culture of the enterococci (a.k.a. **Streptococcus faecalis** group). These bacteria form brownish-black colonies with brown halos.

Plate count agar (peptone glucose yeast agar): This has been a major medium used to enumerate **heterotrophic bacteria** by the spreadplate method.

Potato Dextrose agar (PD): This medium allows a broad spectrum enumeration of both **yeast** and **fungi (molds).** Yeast will tend to form large (3 to 6 mm diameter) colonies which have a domed profile and may be pigmented a pink color. Some yeast emit fruity odors (esters) which are very distinctive. A broad range of fungi will also grow well on this medium.

R2A agar: This is a more complex medium which has been found to support a broad range of **aquatic heterotrophic bacteria**. It is, however, somewhat more difficult to prepare. This medium is being used more widely by spreadplate to determine the total bacterial count and is particularly supportive of many of the **pseudomonad bacteria** which frequently cause problems in controlled recreational water systems (e.g., whirlpools and swimming pools). Some laboratories use dilutions of this medium and claim to obtain greater recoveries (i.e., higher colony counts). Common dilutions are R2A/10 and R2A/100. R2A will also support the growth of some actinomycetes and, in particular, the *Streptomyces* (a common source of the earthy-musty odors associated sometimes with taste and odor problems in water).

Starch-casein agar: This is an agar medium developed for the selective enumeration of **mycelial bacteria (*Actinomycetes*)**. These bacteria resemble molds producing cotton-like or chalky growths like fungi except that here the growths are tightly bonded to the agar and the texture is often leathery. The fluffy appearance is due to the frequent production of raised spore bodies.

Triple sugar iron agar(TSA): This medium was developed for the examination of food products for **enteric bacteria** (Section 5, Family 1) but is applicable to groundwaters. Some ECPS-forming bacteria will produce copious mucoid formations which may fill the void space between the agar and the lid of the petri dish. Useful medium to determine whether enteric bacteria (goes yellow due to lactose fermentation; red if proteolysis occurs; black if hydrogen sulfide is produced) or slime-forming bacteria (excessive mucoid colonies formed) are present and dominant.

Tryptone glucose extract agar: This is a general medium used for the culture of **heterotrophic bacteria**. A wide spectrum of heterotrophic bacteria will grow on this medium which is a minor variation of the standard plate count medium. Here, the yeast extract has been replaced by a slightly higher concentration of beef extract.

Wong's medium: A very selective medium for the culture and enumeration of bacteria belonging to the genus *Klebsiella*. Colonies are pink to red in color and may be somewhat mucoid.

WR agar: This medium is based on the Winogradsky-Regina formulation for culturing **iron related bacteria** (CHI group). This medium generates brown colonies for the typical heterotrophic iron related bacteria due to the uptake of iron from the agar medium. Under circumstances where the groundwater is from regions with a high level of organic pollution (e.g., gasoline or solvent plume), the biodegraders may grow on the WR agar but produce an atypical colony type (commonly white or beige).

Choice of Culture Medium

This is dependent on the intensity of the examination to be undertaken and the perceived potential microbial activities that could be occurring in the groundwater. Typically in an intensive evaluation of an aerobic fouling of groundwater, the Regina concept is to use the following agar media: BHI/4 (general aerobic bacteria); WR (heterotrophic IRB); AA (pseudomonads); R2A/10 (prescreened with 15 mins exposure to 500 ppm of chlorine as sodium hypochlorite for the *Streptomycetes*); TSA (radical slime-forming bacteria and enterics); CZ or PD (fungi and yeast in the case of the PD medium); and LES (coliforms). The selection of a medium for the determination of anaerobic bacteria remains difficult because many of the strictly anaerobic bacteria are both fastidious (require complex media) and sensitive to oxygen (require refined anaerobic dilution and cultural methods). These are difficult to culture in the routine microbiology laboratory unless it has been equipped to handle anaerobes.

Anaerobic incubation of spreadplates prepared "normally" under aerobic conditions will frequently allow the growth of facultative anaerobes and the aerotolerant anaerobes only. A fastidious medium such as BHI/4 will support many of these bacterial types.

Dispersion of Particulates (vortexing)

When a water sample is taken it would be very unusual, but not impossible, for all of the microorganisms to be dispersed in the planktonic phase (separate cells freely suspended in the water). Commonly, a significant proportion of the cells will be entrapped within particulate structures. This would be particularly true where the water has passed through a zone of biofouling and contained a high number of sessile suspended particulates. To maximize the enumeration of potential cfu/mL, it becomes necessary to disperse the particles so that, theoretically each particle would contain a single viable entity which would then each form one colony (cfu). This is clearly a theoretical "ideal". In practice, the water sample can be subjected to vortical agitation using a vibrating horizontal plate which rotates. Such devices are commonly employed in microbiology laboratories to mix dilutions and suspensions evenly. In practice, there is no ideal time to "vortex" a water sample but common practice employs between 10 and 60 seconds, depending upon the clarity of the water sample (clear to turbid, respectively).

Serial (extinction) Dilution of the Water Sample

The objective of a serial dilution is to dilute out the water in a sterile solution (commonly Ringer's isotonic solution) so that there are sufficient viable entities to generate a statistically satisfactory number. For spreadplate enumeration, the usual range is between 30 and 300 colonies if the colony count is to be considered valid. For membrane

filters, the acceptable range is between 20 and 80^{43}.
Commonly the dilutions are made in a tenfold series of
dilutions to (and often beyond) extinction. This extinction
occurs when there are no colonies formed on the spreadplate
due to the dilution being so great that there are no viable
entities in the volume of diluent being streaked onto an agar
culture by the streakplate technique. That means there is a
zero probability of even a single colony growing enumeration.

The technique for conducting a serial dilution and
spreadplate are well described in the literature and will not
be described in detail here. However, each diluent may be
subjected to a streakplate enumeration. Diluents that are
utilized are generally streaked from 0.1 mL of the diluent
although some laboratories prefer to use 0.5 or 1.0 mL. In
the interests of economy, it would appear to be important to
only streak out those dilutions which are likely to have
statistically acceptable counts. There is little point in trying
to count spreadplates which have too many colonies (i.e.,
greater than 300, **too numerous to count, T.N.T.C.**) or too
few colonies (i.e., less than 30, **too few to count,
T.F.T.C.**). Through practice of the art, the appropriate
range of dilutions can be selected which would maximize the
probability of obtaining acceptable colony counts[43,46].

Selection of the dilution range may be initially decided
using the given below. Dilutions are referred to in the
tenfold dilution sequence by the symbol 10^{-x} (ten to the
minus x) where x refers to the incremental dilution. For
example, 10^{-3} dilution would mean three dilution increments
of 10 (i.e., 10 x 10 x 10 or 1,000th dilution). In this
system, x refers to the number of zeros following the one (1)
by which the dilution was expedited. In general, the source
and clarity of the water sample is instrumental on deciding
the number of dilutions that will be performed. In simple
terms, there can be three basic groups: (A) a clear water

without any evidence of excessive biological activity being present; (B) a moderately clear water which may have passed through a zone of biofouling or shows evidence of biological activity (e.g., color, odor, cloudiness); (C) a cloudy or colored water with evidence of fouling (e.g., septic odor, intense color, sediment). Each of these types of water would require different dilutions to be "plated out" by the spread-plate technique if satisfactory colony counts are likely to be obtained.

Water sample group A may be expected to normally have a low bacterial population and the priority for spread-plates can be 10^{-1}, original water and 10^{-3}. Sample group B priority for spreadplating would be 10^{-4}, 10^{-2} and 10^{-1}. Group C water samples would naturally be expected to have much higher populations of microorganisms and greater dilutions should be subjected to spreadplate analysis: 10^{-5}, 10^{-3} and 10^{-1}. It should be noted that in some cases where is there is a larger amount of sediment and rust, very few microorganisms may be recoverable even though it would appear that the water should contain a high population. If there is a septic odor in the water and sewage pollution is suspected for a group C water sample, T.N.T.C. may occur even on the 10^{-5} dilution. In such events, a 10^{-7} should also be included.

Where the inoculum selected for the spreadplate is only 0.1 mL, then the selected dilutions should be reduced by one tenfold increment. For example, a recommended 10^{-3} would be replaced by a 10^{-2} dilution where 0.1 mL was used. Plating out duplications of each dilution will increase the confidence of the data obtained. However, it is generally regarded that at least five replicates of each dilution should be plated out to achieve statistical validity.

Dispensing of Diluent onto Agar Surface (spreadplate).
Three common volumes of diluent may be elected to

form the inoculum for spreading out over the agar surface. These are 1.0, 0.5 and 0.1 mL. The latter two volumes require a correction to the standard formula for computing the population as cfu/mL from the colony counts obtained. For the latter two volumes the correction factor would be times two (x2) and times ten (x10) respectively.

It is important in spreading the diluent over the agar to maximize the coverage of the agar surface with the diluent. This is so that the microbial cells may become dispersed to a maximum extent so that each viable entity has the highest possible opportunity to grow large enough to form a visible colony. The most common technique is to use an L-shaped glass spreader which is sterilized in a flame just before being applied to the droplet of freshly dispensed diluent on the agar surface. Next, the short side of the glass L is gently moved through the droplet and on over the surface of the agar in swirling motions. By this action, the microbial cells are dispersed over the agar surface in a relatively even manner.

Drying the Agar Surface (optional)

Where 1.0 or 0.5 mL of diluent has been used in the preparation of the spreadplate, a significant amount of liquid water may remain on the surface of the agar. Such free water can cause problems during the subsequent incubation through colony growths "running together" to become uncountable and there is also a greater risk of contamination. It is important to ensure that where free water is still evident on the agar surface, it be dried off before the plates are incubated. This can be achieved by tipping the Petri dish lid at an angle to the base so that there is a better movement of air over the surface of the agar to increase evaporation. The lid should, however, still cover the agar surface so that the agar is not directly exposed to any particles falling from the air. Unless conditions are very humid, the drying should be complete within two hours. In these conditions, the smaller inoculum should be used to eliminate the need for drying.

Selection of Incubation Conditions

The next critical phase in the analytical procedure is providing an environment conducive to the formation of colonies of the desired types of organisms where these are present. Three major parameters involved in the selection are temperature, atmosphere and time of incubation before counting the colonies. These selections are critical to the appropriate examination of the water for the aggressive (active) microflora. For example, if a water sample were to be taken from an uncontaminated groundwater source which had a stable temperature of 3.7°C it would not be appropriate to incubate the spreadplates for two days at 35°C. The temperature would be literally toxic to many of the microflora and the few that could flourish at that temperature would not have time to adapt.

Consideration has to be given to the conditions from which the water sample was obtained and the objective of the exercise. If the objective is to determine the potential for microbial biofouling or biodegradation, the incubations conditions should "mimic", as closely as is conveniently possible, the original environment. On the other hand, if the objective is to determine the potential likelihood of hygiene risk through fecal contamination, it may be appropriate to concentrate on environmental conditions known to be favorable to the contaminating organisms (i.e., 35°C with a minimum incubation period of 24 hours).

The former mandate for the indigenous microflora would naturally put a considerable strain on an analytic environmental microbiology laboratory because there would need to be such a variety of temperatures available. It is therefore essential to categorize the water into groups wherein each group is incubated at a different temperature. For spreadplate analysis, the temperature selections automatically link the length of time to the incubation period. It is generally accepted that the higher the temperature over the range from 4 to 37°C, then the shorter the incubation period needs to be. However, the length of the incubation period before all of the capable cfu entities have actually grown to form a visible colony will vary. This variability will primarily be controlled by the length of time that the entity takes to adapt to the cultural conditions presented and assumes that the "territory to be occupied" does not become dominated in the meantime by another colony or colonies. The ideal would be to continue the incubation period until the number of colonies counted is stable (i.e., all cfu entities able to grow have grown to form visible and recordable colonies). In reality this is rarely practicable except as a research exercise.

Two incubation time mandates can be projected for each group separated by the source water temperature. The first time relates to the "natural" fouling or degrading flora while the time for the hygiene risk organisms is discussed later. Recommended incubation conditions are: water group 0 to 15°C, incubate at 12°C for psychrotrophs for 21 days (count weekly), where the temperature is above 10°C additional or alternate incubation at 25°C for 21 days (count weekly) to determine the number of facultative psychrotrophs; water group 16 to 27°C, incubate at 25°C for 14 days (count also at 7 days); water group 28 to 35°C, incubate at 35°C for 7 days (count also at 2 days).

For the hygiene risk organisms (e.g., coliforms and nosocomial pathogens, the general guidelines specify incubation for 24 (extendable on occasions to 48 hours) at 35°C. This is based on the premise that these organisms are able to rapidly adapt and grow under these optimized conditions. In reality, the microorganisms may be under some level of trauma which may require either a more supportive culture medium or prolonged incubation. Such prolonged incubation does not usually extend to beyond seven days. Coliform colonies which appeared after such extended incubation would first of all be confirmed by identification to be *Escherichia coli*. Once this had been established, the water would be considered to contain traumatized coliforms and a hygiene risk may now be established.

Counting Colonies on a Spreadplate

When a viable entity settles onto the agar surface it may be a single planktonic cell, a particulate mass containing only one type of microbe or a mass containing a consortium of several types of microorganism. Each has the potential to produce a single colony. For the latter event, the consortial members would compete for dominance in which the subsequent (mixed) colony formed would contain only the surviving strains which may not reflect the original composition of the consortium. In the two former events, only a single strain of organism was involved and so these colonies are likely to be "pure".

Clearly there would not be time or facility to completely determine which colonies are pure and which are mixed. In the colony count procedure, it is customary to count all of the colonies which are **typical** (i.e., fit the standard characteristics ascribed to the particular microorganisms expected to grow on the medium). **Atypical** colonies are those that do not fit the standard descriptors. Such atypical colonies are generally viewed as being anomalous events reflecting abnormal circumstances.

In some cases all colonies are considered typical (i.e., a total plate count). Enumeration of colonies in a total plate count includes all visible colonies. A number of T.V. image analysis systems are now in common use for such counting procedures. Manual counting is still very commonly practiced. This method generally places the incubated agar plate to be counted on a back illuminated dark ground screen (such as the Quebec counter). The background is dark and the colonies are illuminated from the side. Plates are placed in a rack with the base towards the observer. As each colony is counted a small mark is made (e.g., dot with a permanent marker) on the base of the dish. The technician records on a hand counter, electronic digitizer or remembers the count to record at the end. Where there are more than a hundred colonies on the plate, the base can be divided into four quadrants and only one quadrant of colonies counted. The number of colonies counted is then multiplied by four to give the total colony count for the plate.

Some selective media do differentiate out colony types as being typical of a particular group of organisms. Where these media are used, the colonies should be counted separately by typical group as specified in the media and standard methods documentation. Other colonies may be recorded as atypical. On some occasions, these atypical colony forms may be so numerous that they cover the agar surface and mask any typical colonial growths that may have been occurring. This condition is referred to as an **overgrowth**. Greater dilutions can sometimes be used to resolve such overgrowth events but, in cases where the atypical cfu exceeds the typical cfu forms, extended dilution and spreadplating does not resolve the issue.

Computing the cfu/mL or cfu/100mL

Data has now been gathered on the numbers of colonies recorded forming on the agar spreadplate at the various dilutions (as total plate, typical or atypical counts)[43,46]. The simplest interpretation is where there has been no colonial growth. This would normally be expressed as 0 cfu/mL or 0 cfu/100mL whichever would be appropriate. Unfortunately, such a result would suggest that none of microbial types under investigation were present whereas, in reality, this result indicates that these microbial types were **not detected (N/D)**. A number of factors (e.g., trauma, competition, inappropriate sampling, dilution and cultural conditions) could have prevented colony expression. Care must therefore be practised in the interpretation of the results and consideration should be given to the probability that the resulting data would be of **comparative** value rather than of **absolute** value. It is interesting to note that in the medical industry where microbial presences are likely to be associated with specific microbes (i.e., the pathogen), enumerations tend to be given an absolute value (i.e., the pathogen is present in these numbers or, in absence, these microorganisms did not cause the disease). In groundwater there is a complex community structure which reduces the value of the data to a comparative level. Exceptions would include anomalous conditions such as where a major aerobic biodegradation was occurring. Here, the organisms associated with the degradation events would dominate and a more precise enumeration may subsequently become achievable.

There are two primary mechanisms by which the water sample may be enumerated using the agar spreadplate technique. One method primarily examines expressions of colonial growth from dilutions of the water sample (streaked spreadplate). The second method examines various volumes of the water sample itself to determine whether microorganisms are present through their entrapment and subsequent growth on a filter (membrane filtration technique). High population numbers (cfu/mL) are easiest recorded using the streaked spreadplate while low population numbers (cfu/-100mL) are more conveniently enumerated using the membrane filtration technique. Each methodology involves a different manner of calculating the population from the numbers of colonies recorded. These will be discussed below.

Streaked Spreadplate (calculation formula)

A series of colony counts (CC_d) are obtained for each of

the dilutions (d). Where there has been an inadequate dilution of the water, the number of colonies may be **too numerous to count (T.N.T.C.)**. This is generally considered to be where more than 300 colonies can be enumerated by direct visual examination. If the dilutions are too high then the number of colonies would be **too few to count (T.F.T.C.)**. This may be considered to have happened where there are fewer than 30 enumerable colonies on a given plate. There remains the possibility that two tenfold dilutions may yield 301 colonies at one dilution and only 29 colonies at the next incremental dilution rendering it technically impossible to ascribe a count. There is therefore some flexibility built into the interpretation of these spreadplates. The ideal methodology would be to use a replicate analysis which should minimally include six replicates for each dilution and ideally incorporate fifteen. Clearly, this is not practicable for economic reasons.

Midpoint cfu/mL formulation: One formulation for obtaining the cfu/mL involves the selection of the dilution streaked spreadplate which displays between 100 to 150 colonies. This is at the midpoint of the acceptable range for computing the cfu/mL. Where all other colony counts are outside the range (i.e., 30 to 300), the population may be calculated using the following formula:

$$cfu/mL = CC_d \times Y$$

where CC_d is the colony count at dilution d and Y is the multiplier factor based upon 1 followed by a series of zeros equivalent to the tenfold dilution factor d. For example, where d is 10^{-4}, then factor Y would be 1, followed by four zeros (0000) to make the number 10,000. Where the CC_d was 125, the cfu/mL would be 125 x 10,000 or 1,250,000 cfu/mL.

Coupled cfu/mL formulation: Where there are more than one CC_d falling within the range of 30 to 300 colonies on the streaked spreadplate, the calculation has to be modified to allow an equal weighting to each of the dilutions in the range. Calculate the cfu/mL$_d$ for each dilution d based on the equation:

$$cfu/mL_d \quad = \quad CC_d \; x \; Y_d$$

Once the cfu/mL$_d$ have been calculated for each relevant colony count then the final cfu/mL may be calculated as the mean value of the sum of the cfu/mL$_d$ values obtained:

$$cfu/mL \quad = \quad (\text{Sum of Z cfu/mL}_d \text{ values}) \, / \, Z$$

where Z is the number of cfu/mL$_d$ values used.

Projected cfu/mL formulation: If there are no colony counts on the streaked spreadplates but there are counts higher and lower than the acceptable range, the cfu/mL may be projected on the basis of the cfu/mL$_d$ data obtained closest to the acceptable range (i.e., 15 to 29 and 301 to 500). Here, the cfu/mL would be calculated in the same manner as the coupled formulation given immediately above.

Fluctuating cfu/mL formulation: Where there is a con-siderable amount of fouled material (e.g., sloughed biofilm, soil particles), it is quite probable that the CC$_d$ values may fluctuate radically down the dilution scale. Spurious high colony counts in the greater dilutions may be a result of the breakup of particulate structures which then generate a larger number of colony forming units. Secondary releases of particles (and cells) which had become attached to walls of the diluting equipment and had later released in much higher dilutions may also cause this effect. The net effect is that the CC$_d$ values decline more slowly through the acceptable range and then, at greater dilutions suddenly exhibit higher CC$_d$ values. Streaked spreadplate enumerations from soils are sometimes very prone to these "quirks". It has been known for dilutions as high as 10^{-10} to 10^{-15} to display abnormally high colony counts for the reasons given above. When this event occurs, the cfu/mL should be calculated using the midpoint cfu/mL formulation taking the CC$_d$ data for the first dilution in the series which falls to within the acceptable range. It should also be noted that where the data is radically fluctuating this could indicate a major biofouling event or a significant presence of particles.

Spreadplate enumeration of bacterial loadings (cfu/mL), projected populations of specific groups of bacteria can be determined by culturing various dilutions of the water sample on different agar media. Dilutions recommended are 10^{-1}, 10^{-2} and 10^{-4} in sterile Ringer's solution. Volume spread per plate may be 0.1 mL (high humidity) or 1.0 mL (low humidity). Incubation temperatures have to be related to the original temperature of the water being pumped from the water well. It is perhaps most convenient to utilize room temperature (20 to 24°C) where the water temperature ranges from 12 to 30°C. Incubation times should be long enough to allow the bacterial colonies to develop; this occurs in a stepwise series of increases. It is recommended that the agar plates be enumerated after 14 days. However, if there is urgency in obtaining the data, then 7 days incubation can be substituted, recognizing that a lower number of colonies may be counted. If there is time, prolonged incubation for 4 to 6 weeks may reveal a higher colony count. Data will be presented as <u>colony-forming units per mL</u> and the specific group of bacteria being enumerated will be dependent upon the type of agar culture medium being used. The following is a list of agar culture media which can be used to determine specific groups of bacteria:

HB	R2A agar	- aerobic	- <u>Heterotrophic bacteria</u>
IRB	WR agar	- aerobic	- <u>Iron-related bacteria</u>
GAB	BHI/4 agar*	- aerobic	- <u>Gross aerobic bacteria</u>
SRB	Postgates B	- anaerobic	- <u>Sulfate-reducing bacteria</u>
PSE	AA agar	- aerobic	- <u>Pseudomonads</u>
STR	R2A/10 agar*	- aerobic#	- <u>Streptomycetes</u>
MLC	Czapek-Dox agar	- aerobic	- <u>Molds</u>
MLP	PDA agar*	- aerobic	- <u>Molds</u>
CLF	M Endo LES agar	- aerobic@	- <u>Coliforms</u>

The agar culture media marked with an asterisk (*) possess some unique features; BHI/4 is a brain heart infusion agar at 25% of the nutrient loading; R2A/10 is the R2A medium at 10% of the normal nutrient loading and the water sample is pretreated with 100 ppm of sodium hypochlorite (subsequently neutralized with sodium thiosulfate #) to reduce the bacterial vegetative cell count and allow the exospores of *Streptomycetes species* to survive and grow on the modified R2A agar; PDA agar is potato dextrose agar which supports the growth of a wide variety of molds; and (@) the M Endo LES agar is incubated at 35°C for 24 hours only to enumerate coliform bacteria.

Spreadplate bacterial population relationships (SPB-R), as water is pumped from a larger and larger zone of influence, so there are changes in the microbial composition of the water as a result of the "fringe biozone" organisms arriving at the well. One manner in which these shifts may be observed is to examine the interrelationships of the various bacterial populations. This may be achieved by comparing the populations as SPBR factorials. Here, the SPBR is determined by the following formula:

SPBR X factor = (Population^2) / (Population^1)

where the ratio would be population^1 : population^2 and expressed as 1 : X, X being the computed factor given as SPBR in the above equation. For example, where the IRB:GAB ratio is to be determined based on respective populations (cfu/mL) of 5,000 and 15,000 the SPBR X factor would be computed as 3.0 and the IRB:GAB ratio would be 1:3.0. While all possible combinations could be compared in this manner, the dominant comparisons currently envisaged are listed below.

IRB:GAB, the largest numbers of IRB are commonly found in the biozones closely associated with the water well installation itself. The ratios in these zones will vary from 1:1 to 1:10 but further away from the well itself, the ratio may shift to 1: > 100. Where water samples are taken from a producing water well, the IRB:GAB ratio may remain relatively constant and reflect, to some extent, the dominance of IRB in the biofouling event (i.e. 1:1 IRB dominant; 1:100 IRB suppressed).

GAB:CLF, in cases where the coliforms are absent from the water sample, this ratio could not be computed. However where a CLF population is recovered, there exists a hygiene risk as well as a biofouling risk. The higher the SPBR X factor (as it approaches 1.0) the greater the level of concern that there has either been: (1) an intrusion of fecal-rich wastewaters; or (2) a substantial growth of some coliform-like bacterium within at least one of the biozones around the well. Concern should be expressed where CLF are recovered and a standard series of coliform tests should be undertaken to confirm the event. Identification of the dominant coliforms should reveal *Escherichia coli* if event (1) has occurred, or *Enterobacter* and/or *Klebsiella species* in the case of event (2).

GAB:PSE, where there is an aerobic degradation of a specific group of organic pollutants (e.g. gasoline, organic solvents) pseudomonads (PSE) may tend to dominate over other bacterial types. This event will cause the GAB:PSE to bias from a GAB dominance (i.e. 1: < 1) to a PSE dominance (i.e. 1: > 1). It should be noted that where there are extremely high pseudomonad populations associated with the degradation of specific organics, there may be very large numbers of atypical colonies (not brown, usually white or cream colored) growing on the WR spreadplates.

GAB:MLC, molds may dominate in groundwaters which include a fraction of water from the capillary and/or unsaturated zones, particularly where these zones have become heavily contaminated with organic wastes (e.g. recharges from oxidation ponds around a pulp mill). Such waters can carry very high populations of mold (fungal) exospores which germinate and grow very readily on the Czapek-Dox (MLC) or potato dextrose agar (MLP). In a saturated groundwater system, the GAB:MLC ratio may not be computable due to the absence of these exospores. However, when such contaminations do occur the GAB:MLC ratio can switch dramatically from 1: <0.001 to 1: >1. Because of the abundance of exospores, it is not unusual to see the SPBR shift to 1: >1,000 in events where there has been a massive mold intrusion. The same conclusions can be drawn from the GAB:MLP ratio.

The composition of the bacterial flora within a bacterial interface around a water well will vary depending upon a number of environmental factors. These can be summarized (Figure 37) to describe the major potential events: (1) aerobic degradation often involving a narrow range of pollutant chemicals and commonly dominated by Pseudomonas species(PSE); (2) a mixed aerobic-anaerobic degradation often involving a broad range of pollutants and a wider spectrum of bacterial types (GAB); (3) a high population of molds (MLC) would indicate that the water may have passed through an unsaturated zone where there was a considerable amount of degradation of organics occurring; (4) where the coliforms (CLF) are found to be a significant part of the flora then there is a serious potential for fecal material to be present and an enhanced hygiene risk; and (5) iron related bacteria (IRB) may be found to be a major (>10%) part of the microflora; there is a strong risk of plugging and/or corrosion.

Other ratios may also be computed where differences in the water characteristics from the different biozones is observed. Unfortunately, the SRB group of bacteria tend to live within tight consortia with other bacteria (often the pseudomonads) and the SPBR is difficult to apply to this group of bacteria.

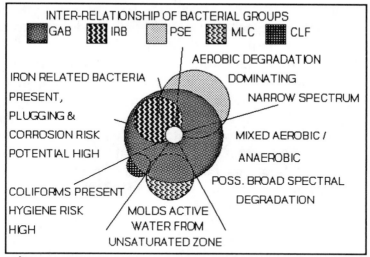

Figure 37. *Schematic of the inter-relationships between some of the major bacterial groups. Examine composition of bacterial flora and match the relationship from the center zone (white circle) to the edge.*

Membrane Filter (calculation formula):

Generally the membrane filter functions where there is expected to be a low number of targeted microorganisms which would be enumerated in the cfu/100 mL range. For membrane filters, the surface area to be enumerated is smaller and the number of colonies for an acceptable CC_v is between 20 and 80 colonies. It must be remembered that volumes (v in mL) are passed through the membrane filter and that the population is enumerated on the basis of the filtered volume of the original sample. That sample can be diluted into sterile isotonic solutions in order to obtain counts where there is a higher density of microorganisms[43].

There is a major interference factor which can distort the value of data obtained by this means. That is the event where the filter not only entraps the microorganisms within its porous structures (normally 0.45 or 0.22 microns) but also inanimate particulate structures. Some of these colloidal particles will "squeeze" through the pores to pass on into the filtrate. Other particles will entrap on the surface of the filter. This would have a number of impacts on the subsequent colonies that grow. These are: (1) shifting in the composition of available nutrients to include those bio-entrapped in the colloids; (2) protecting any incumbent viable entities from any inhibitors present in the selective culture medium; and (3) the colloidal "burden" on the surface of the filter may bias the forms of colonial expression that are likely to occur. One net effect of these interferences is that microorganisms (e.g., pseudomonad bacteria) not commonly expected to grow may flourish and cause **overgrowths** which render the count of typical colonies difficult if not impossible.

Enumeration of the CC_v follows a similar mandate to the CC_d for the streaked spreadplate. Calculation of the population uses a baseline value of 100 mL for the reference volume on most occasions. The same form of computation as the spreadplate can be used with two exceptions: the acceptable CC_v range is now 20 to 80; and the calculation often does not involve dilutions and relates back to 100 mL. An example of the computation of a calculation of the cfu/100 mL is given below:

$$MF_c = 100 / v$$

where MF_c is the correction factor and the CC_v is the colony count falling within the range of 20 to 80 for the volume (v) which would be given as mL. From these data, the populations can be computed using the equation:

$$cfu/100 \text{ mL} = CC_v \times MF_c$$

where the CC_v falls between 20 and 80 colonies and the MF_c is the appropriate correction factor. Because the range of acceptable colony counts is so narrow (i.e., less than a tenfold scale), there is lower likelihood of obtaining two colony counts from within the range. There is a greater possibility, however, that no colony count will fall within the acceptable range. In such an event, the counts from 81 to 160 should be selected.

While the numbers produced by either the spreadplate or the MF techniques would appear to be valid and can extend to three decimals of accuracy, the user should remember that there are a number of limitations. These counts reflect only a part of the total microflora and, unless subjected to rigorous replication, can be inaccurate. When used in a comparative manner, these tests can present a reasonable level of certainty that a particular microbially driven event is occurring.

COLIFORM TESTING

Coliform bacteria includes a number of bacterial genera within the Section 5 of *Bergey's Manual of Systematic Microbiology*. These are grouped into a number of categories:

> Fecal coliform - *Escherichia coli*
> Coliform group - *Klebsiella pneumoniae*
> *Enterobacter aerogenes*

These bacteria are defined as being facultatively anaerobic, gram negative, non-sporing, rod shaped bacteria that ferment lactose with gas production within 48 hours at 35°C. Note that the reason for extending the incubation period to 48 hrs where the tests have been negative at 24 hours is to cover those coliforms where there is a delay for whatever reason before gas is produced. These bacteria make up 10% of the microflora in the large intestine. *Escherichia coli* is universally present while *Klebsiella*[34] and *Enterobacter* species are commonly present (in 40 to 80% of the incidences recorded). However, the coliforms include a range of species which may thrive in various soil and water environments (e.g., sediments). These can generate false positives. Where the fecal coliform test is operated at a higher temperature (44.5°C) there is a restriction of fermentative activity to only the fecal coliforms (*Escherichia coli*). The widespread acceptance of the coliform presence as an indicator of hygiene risk (fecal contamination) is flawed by the fact that these bacteria lose their viability in water[17].

There are two principal quantitative test methods for the coliforms. These are the multiple tube and the membrane filtration methods.

The multiple tube method (such as lactose/lauryl tryptose broth) involves a multiplicity of various dilutions of the water sample in fermentation tubes in a logical sequence which allows a mathematical projection of the most probable number (MPN) where only some of the tubes generate a positive reaction (i.e., gas in a Durham's tube). At the beginning of the twentieth century, Durham developed an inverted glass test tube in the liquid culture medium as a means to entrap and observe the production of gas by microorganisms. It was found that bacteria could be partially identified by the sugars they could break down (to acid with or without gas). Coliforms were found unique in their common ability to ferment lactose to both acid and gas.

Following a presumptive test such as the multiple tube method, there is an optional second phase to confirm the presence of *E. coli* by fermentation in brilliant green bile lactose broth with subsequent (more complete testing) on Levines EMB or endo agar. This test (MPN) is expensive to perform due to the many tubes that have to be employed involving a high level of technician management.

The membrane filter (MF) technique is conceptually simpler, involving the entrapment of any bacterial cells and suspended particles on the upper surface of a fine porous filter membrane (0.45 micron mean pore size). Theoretically, any volume of water could be passaged through the filter and the incumbent cells entrapped. When the filter is transferred to an appropriate medium and incubated, the number of colonies formed can be back-calculated to the population size in the original water sample. The procedures in the MF test method do, on occasion, cause stress in the filtered cells which leads to delays before colony formation occurs. A variety of holding media have been suggested to reduce the stress on the coliforms and improve the potential for complete recovery of all coliform bacteria in a water sample. It is more economical than the MPN test but is thought to be subject to a higher risk of false negatives where coliform bacteria are in stress.

To compensate for the economic costs and relative lack of precision in the MPN and MF methodologies, a simpler test has been developed based on the presence-absence (P/A) of coliforms in the water sample. The medium used is less selective for the coliforms and does provide an environment enriching for other genera of bacteria also sometimes associated with the gastro-enteric tract[6,10,31,43]. These genera include (% range of incidence):

Enterobacter species	40 - 80%
Klebsiella species	40 - 80%
Proteus	5 - 55%
Staphylococcus	30 - 50%
Pseudomonas	3 - 11%
Fecal Streptococci	100%
Clostridium	5 - 35%

This P/A test uses a broth culture comprised of lactose broth, lauryl tryptose broth with bromocresol purple as a pH indicator. It is usually performed using a 100 mL water sample which dilutes the medium to the correct strength. The generation of yellow color indicates that fermentation of the lactose has occurred which is considered to be a positive acidific reaction indicating that bacteria are present which may be linked to fecal contamination. Where foaming is detected, gas production may be considered to have occurred (usually this occurs between 24 and 48 hours into the incubation period at 35°C). "This simple test is gaining acceptance around the world and is a standard procedure in some countries" (Prescott, Harley and Klein, 1990, *Microbiology*, W.C. Brown, Dubuque, IA)[62].

There are a range of coliform tests available using the MF, MPN and the MUG concepts. The latest coliform tests to be approved are based on the MUG system (e.g., Colilert). Below is a description of the new (patent pending) gas thimble (GT) technology which allows very convenient coliform testing in remote and undeveloped facilities.

There are two different COLI-MOR™ test devices which have been designed to detect:

A. Any coliform type of bacteria (those which are able to ferment lactose and produce insoluble gaseous products), are able to grow at 35 to 37°C and will also produce acidic products in a maximum of 24 hours. This is a broad spectrum test designed to investigate the gross coliform group and would have a tendency to produce false positives. There would be a high probability that a negative test (i.e., no acid or gas production) would indicate that the water did not contain any coliform related bacteria but it does not indicate that the water is free from bacterial presences of any type. It parallels the standard presumptive coliform test which functions mainly at the presence/absence level. The standard test operates using acid production as being indicative of a positive reaction while the gross COLI-MOR™ test relies on the generation of gas (see Figure 38). It should be noted that the gas thimble can react **before** there are any signals of acid generation.

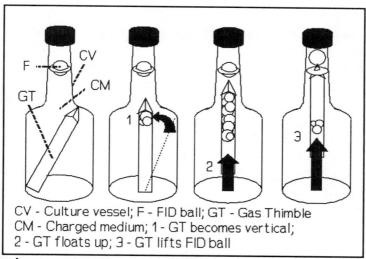

CV - Culture vessel; F - FID ball; GT - Gas Thimble
CM - Charged medium; 1 - GT becomes vertical;
2 - GT floats up; 3 - GT lifts FID ball

Figure 38. *Form of a positive reaction in the COLI-MOR™ (patent pending) coliform test. The semipermeable gas thimble (GT) fills with gas, drops in density and rises (2 and 4).*

The test is known as the:

G-COLI-MOR™

This is recommended as the frontline test for the routine detection of any coliform related bacteria. For water treatment operators, the G-COLI-MOR™ offers an in-house test system for the routine determinations for the presences of the broad spectrum of coliform bacteria as an indicator of the hygiene risk. A consistently negative series of tests would act as a good indicator that the formal standard presumptive coliform test would also be negative.

B. Where there is a need to be more precise in the monitoring for coliform bacteria because the water is primarily potable or could be used in clinical situations where there are immunologically impaired persons, there is a need to determine more precisely the nature of any coliforms which may be present in the water. The objective of this test is to detect the bacteria which are able to ferment lactose with copious gas production under conditions where there are high concentrations of bile salts (which inhibit many of the non-enteric coliform type of bacteria). The medium selected for this test is Brilliant Green Bile Broth which is well acknowledged to be very selective for *Escherichia coli* and related species. The temperature used is 35 to 37°C and the incubation period, 24 hours. The only signal of a positive reaction is a generation of gas which causes the gas thimble to rise above the surface of the medium. This test is known as the:

S-COLI-MOR™

This test may be recommended for potable water supplies and, in particular, water supplies being delivered to hospitals, indoor aquatic recreational facilities, and the food and pharmaceutical industries. The lag time (hours of delay, *hd*) can be used to project the population size of total coliforms present in the water sample using the COLI-SOFT software program. Where positive reactions occur, the standard enrichment techniques can be applied to confirm the nature of the coliforms present in the sample.

Under some conditions, it becomes important to determine the presence and size of the population of *Escherichia coli* in water. There are two additional factors which have been applied to further restrict the diversity of coliform bacteria to just *E. coli*. The medium used for the test is a very selective medium for this genus of bacteria and the temperature of incubation is raised to 44.5°C and the period restricted to 24 hours. A positive reaction is the gas thimble rising to and possibly breaking through the surface of the medium due to gas buildup. The *hd* time can be used to compute the probable population of *E. coli* in the water sample.

Sensitivity
From the laboratory testing to date, the theoretical range of sensitivity of the tests to the target bacteria is one cell in ten liters of water. Less attention is now directed to chemical signals (e.g., nitrate as representing a fecal pollution risk[38]). In practice, the limit is restricted by the size of the water sample which is restricted to 100 mL. The normal reactivity measured as *hd* ranges from as short as seven hours for a population of 10^8 coliforms/100 mL to nineteen hours where there is a single target cell present in the 100 mL sample. There is an inverse log linear relationship (R, 0.96 for ATCC strains) between the population size and the *hd* (hours of delay to positivity).

Disposal
These tests, when completed, offer a potential health risk due to the possible active growth or presence of pathogenic bacteria. Careful disposal of completed tests is recommended by following the manufacturer's guidelines.

In the Event of Spillage
Accidents happen which can involve the spillage of the contents of an active or concluded tests on the bench, desk

or floor. When this occurs, care should be taken to thoroughly disinfect the areas affected. For solid plastic surfaces, the affected area should be flooded with either a bleach or strongly bactericidal solution (such as a quaternary ammonium compound), left for ten minutes and then mopped up. All swabbing materials should be bagged in plastic bags and disposed while the operator washes hands and arms with a recognized disinfectant soap. For solid permeable surfaces (e.g., untreated wood, carpet or cardboard), the length of contact time should be increased from ten to thirty minutes. It is recommended that a white laboratory coat be worn during the operation of the tests and that all recognized microbiology laboratory procedures be followed[43]. When the coat does get contaminated, carefully remove the coat and treat all affected areas with a disinfectant for thirty minutes. Afterwards these areas should be rinsed with water and the coat, after treatment, sent for laundering.

The Components of the MOR™ Test
Each COLI-MOR™ test consists of two basic components (see video presentation disk B, SHOW COLI). These are:

COLI - culture vessel

Gas thimble tube

The COLI-culture vessel appears to be shaped like a Worcestershire sauce bottle and contains 50 mL of triple strength culture medium recommended for the particular test. The neck of the culture bottle acts as a guide for the gas thimble when it is rising to the surface. The expiry date for the test is given on this label. There is an upper label the top of which forms the fill line for the test. When the water sample has been added to this line, 100 mL of water would have been added and the medium diluted to normal strength. This upper label also carries a scale to measure the height (in millimeters) of any elevation in the cultured medium during incubation (due to displacement by gas generation). The gas thimble tube consists of a disposable tube holding a sterile gas thimble in which the open end (of the thimble) is facing towards the open end of the containing tube. This thimble is made of a sterile weave material with a maximum width of 9 mm. The advantage of the gas thimble tube over the Durham's is: (1) microorganisms and chemicals are able to move between the tube and the main body of the liquid culture medium; and (2) the tube protects two visible signals where gas is produced (i.e., elevation of liquid level, and upward relocation of the thimble itself).

Above the thimble is a patent-pending floating intercedent device (FID) which restricts oxygen entry into the liquid charged medium. The presence of oxygen may delay the reactivity of the system.

SETTING UP A COLI-MOR™ TEST

1. Obtain the water sample using the standard methods.
2. Label the lower label of the COLI- vessel with the sample origin, date and time in the spaces allocated.
3. Unscrew cap and lay on clean surface without turning over (to prevent contamination of the inside surfaces).
4. Pour sample water into the vessel until the liquid level rises to the fill line on the upper label. Do not directly touch the inner surfaces of the vessel since contamination could result.
5. Take the gas thimble tube and remove the cap while holding the tube upright. Quickly bring the mouth of the tube over the mouth of the vessel and invert the tube. This will cause the thimble and FID to slide from the tube into the vessel. The gas thimble will settle in the liquid.
6. Recap the vessel tightly.
7. The COLI-MOR™ test will require incubation in the upright position at 34 to 38°C for either the G- or S- series.

8. After 24 hours of incubation, the visible occurrence of the gas thimble rising up to or above the liquid medium level is to be taken as a positive indication of the presence of coliforms in the sample.

G-COLI-MOR™ Modification of the P/A test in which there are two levels of reactivity:

Positive One: Medium changes from a clear red to a dirty yellow color:
Bacterial genera which could be present:
> *Escherichia coli*
> *Klebsiella pneumoniae*
> *Enterobacter aerogenes*
> *Aeromonas*
> *Staphylococcus*
> *Pseudomonas*
> *Fecal Streptococci*
> *Clostridium*

Positive Two: Gas thimble rises:
> *Escherichia coli*
> *Klebsiella pneumoniae*
> *Enterobacter aerogenes*

This test operates therefore at two levels of sensitivity and it should be noted that one of the positive reactions may occur without the occurrence of the other positive reaction. A positive one reaction may be considered to indicate that bacteria of fecal origin may be present in the water sample and confirmation would be required. A positive two reaction would indicate that coliforms are present in the water sample. The broad spectrum of bacteria which may generate a reaction in this has led to the test being designated G- for the gross range of bacterial presences which may be recorded.

S-COLI-MOR™ Confirmation of the presence of *Escherichia coli* employs more selective media. One of the most commonly employed selective culture media is brilliant green lactose bile broth. This medium does not employ a color change but does provide restrictors to growth in the form of lactose (sole source of carbon), high concentrations of bile salts (to inhibit nonenteric bacteria) and brilliant green (selective inhibitor). Gas generation is the sole source of a positive reaction and the incubation time is 24 hours at 35°C.

Positive One: Gas thimble rises to a visible position:

> *Escherichia coli*
> *Klebsiella pneumoniae*
> *Enterobacter aerogenes*

This level of testing is more specific for the coliforms as such and hence it is referred to as a specific (S-) test.

ENUMERATION BY SEQUENTIAL DILUTION TECHNIQUES

Over the last fifty years, microbiologists have tended to rely mainly upon the use of agar culture media in the enumeration and primary identification of microbial agents. There have, however, been periodic initiatives to enumerate or determine microbial activity in incremental dilution series using liquid culture media as both the diluent and culture medium. Two approaches to the determination of microbial numbers and/or activities are commonly practiced. These are the **tenfold serial dilution technique** and the statistically validated **most probable number (MPN)**. The former method involves the logical dilution of the water sample to beyond the realm where recoverable viable entities could be present. In the latter technique, various volumes of water are investigated to determine which aliquots still contain demonstrable microbial presence. Replicates of the volumes selected are used to generate a statistical determination of the microbial population.

The additional use of a diluent as the culture medium for growing the targeted microorganisms replicates perhaps more closely the natural environment in which these organisms usually flourish than the surface of an agar culture medium. It is naturally assumed that the density of the cells declines in a direct relationship to the degree of the dilution applied. For example, a water sample containing 3,300 cfu/mL would be expected upon a hundredfold dilution to contain 33 cfu/mL. A further hundredfold dilution of the dilution would theoretically contain 0.33 cfu/mL or it could be expected that there would be a one in three (1 out of 3) probability that any given mL of the diluent would contain a viable entity.

Application of the tenfold serial dilution technique is most likely to generate data which would indicate a differential point where one dilution (less dilute) would be positive while the next dilution (more dilute) would be negative. It can be extrapolated that the population must be greater than the critical population (i.e., at least one viable entity in the total volume of the diluent) than the dilution factor applied to the original water sample since growth had occurred. At the same time the population must be smaller than would allow growth to occur in the next higher tenfold dilution. Generally therefore, the population expressed by this technique would be given as greater than ($>$) the highest dilution factor showing growth. For example, in a tenfold series down to 10^{-6}, growth was recorded down to 10^{-3} dilution but not beyond. It can be extrapolated that the population is minimally 10^3 viable units per mL or 1,000 v.u./mL. This would be expressed as $> 1,000$ v.u./mL. At the same time, the population would have to be considered as being less than 10,000 v.u./mL; otherwise the next highest dilution would also have exhibited growth. The tenfold serial dilution may therefore be considered to operate more at the semi-quantitative level of accuracy.

Most probable number (MPN) technology is more applicable to low populations of microorganisms which may be associable with a hygiene risk concern. Most commonly the MPN is used for the evaluation of coliforms in water where these bacteria may be present in numbers between 1 and 180 per 100 mL. Typical water sample volumes to be employed are: Scenario A, 5 tubes each containing 10 mL of water plus the medium; Scenario B, 10 tubes each containing 10 mL of water plus the medium; and Scenario C, 5 tubes of each of 10 mL, 1 mL and 0.1 mL of water plus the medium are used. In all three cases the number of positives is used to determine the MPN index (likely number of coliform organisms per 100 mL) and the 95% confidence lower and upper limits. Nineteen times out of twenty the number of coliforms present in the water sample will fall within these upper and lower confidence limits. Positives are taken to be those tests which show fermentation (visible gas formation) within the incubation period (usually 24 hrs) at 35°C.

The MPN technology does have application to investigations where there is expected to be: (1) a low number of targeted organisms in the water; (2) a large potential population of other microorganisms which could interfere with enumeration procedures; (3) a selective medium or specific product display which will signal activity generated solely by the targeted organisms; and (4) incubation conditions which will suppress the non-target flora.

Extinction Dilution Biodetection (EDB) scenario. It is sometimes necessary to determine the microbial activity levels within porous media within a sediment, soil or aquifer system, particularly in a comparative manner (e.g., to determine the differences between "pristine" and (pollutant) impacted media)[9,24,44,52,60]. Traditional technologies using the dilution/spreadplate methodologies can become time-consuming and difficult to interpret. An alternative technique is to use the extinction dilution\BART™ biodetection (EDB) scenario. Here, the samples of porous media are taken from the potentially (pollutant) impacted area(s) together with pristine samples that can be reasonably considered to have come from a comparable formation which has not been impacted upon by the pollutant event. The rate at which diluents of the porous media trigger both *dd* and RX events will, in a comparative manner, allow the relative aggressivity and population dominance to be determined. For example, in an aerobic unsaturated porous medium it may be expected that the microflora may be dominated by aerobic sessile bacteria which would trigger a short *dd* on a SLYM-, FLOR- or TAB- BART™. At the same time, the high aerobicity would reduce the likelihood of such an anaerobic event as the generation of an aggressive SRB population. Where a pollution event (e.g., gasoline intrusion) occurs under such conditions, the increased level of biological activity would remove much of the oxygen to cause anaerobic conditions. Under these circumstances, the SRB component of the microflora may flourish as a direct result of the increased level of biofouling now occurring. In this event, it may be expected that the *dd* for SRB-BART™ may become shorter (i.e., a greater level of activity) while the *dd* for the SLYM-, FLOR- and/or TAB-BART™ would lengthen. A quantitative measure of the population size may be obtained by observing the (tenfold) level of dilution at which a *dd* is still obtained. Greater dilutions may be considered not to have had a sufficient population base to allow growth to occur within the biodetector during the period of observation.

A recommended scenario for undertaking an EDB study would be to minimally have two samples of the porous medium under question. One (Z) would be from a zone impacted upon by the event under investigation. The second sample would be from an area within the porous medium which may reasonably be considered to be outside the zone of impact and yet, in other ways, similar to the porous medium obtained as sample Z. This second sample forms the background (B) or control conditions which would be occurring in a non-impacted state. A number of stages need to be followed in the microbiologic investigation of these samples after the successful acquisition of samples.

These stages include: (1) observation of the (soil, core) sample before the laboratory investigation for any visible or odor changes which may be associable with microbial activities; (2) sorting of samples into groups based upon differences seen in stage 1; (3) collection of a number of samples from each group which can be blended together to form the composite sample to be investigated; (4) a tenfold dilution of each composite subsample into sterile isotonic (e.g., Ringer's) solutions beginning with a 10 mL (or 10 g) sample dispersed into 90 mL of sterile diluent to form a 10^{-1} dilution; (5) vortical dispersion of the diluted composite sample to obtain an even distribution of the particulate material; (6) allow the denser particulates to settle out; (7) conduct a tenfold serial dilution of the suspension in sterile

Ringer's; (8) aseptically dispense 1 mL aliquots of each diluent into designated precharged (14 mL sterile Ringer's solution per biodetector) BART™ tubes; (9) incubate the charged biodetectors at an appropriate temperature and observe daily for the initiation of a recordable reaction (as RX) in days of delay (*dd*); (10) continue observations and record any shifts in the reaction pattern (RX2, RX3) which may occur; and (10) compare the data obtained for the Z sample(s) with that for the B sample(s).

Relative population sizes may be computed using the *dd* data obtained over the fourteen day observation period. The methodology for obtaining the relative population estimates is to use the *dd* data from the highest (most) dilute increment in the series which gives a positive reaction by the end of the observation period. The possible population (*pp*) is automatically projected if the data have been filed in the BART™-SOFT for all positive *dd* as these occur. The population size may be projected by the following method:

Most dilute tenfold increment displaying a *dd*: 10^{-X}

Serial dilution increment: X

Possible population projected (cfu/mL): PP_X

From these data the incumbent possible population (**IPP** cfu/mL porous medium) may be calculated as:

IPP = PP_X * Antilog of X.

The antilog of X may be calculated as a one (1) followed by the same number of zeros (0) as given in the X value. For example where X equals 3 then the antilog would be 1,000.

For each of the different bacterial groups included, a factorial comparison (FC) of the IPP values (e.g., IPP_{irb}, IPP_{srb} and IPP_{slym}) may be made using the formula:

$FC_? = IPPZ_? / IPPB_?$

where ? is the group of bacteria being compared, IPP data is presented for the Z sample and the comparable B sample.

An anaerobic environment may be considered to have been generated in the impacted soil where the $FC_?$ is greater than 1.3 for the SRB-, and IRB- bacterial groups and less than 0.7 for the SLYM-, and FLOR- bacterial groups.

Population shifts that occur between the B (control) sample(s) and the Z (impacted) sample(s) may be calculated as factorial relationships (FR) of the dd_z to the dd_b values at the greatest dilution at which both dilutions have registered a positive dd:

$$FR = dd_z / dd_b.$$

Where the FR value is > 1.0 then that particular bacterial group may be considered to be more aggressive in the impacted (Z) porous medium than the control (B) medium. Conversely, where the FR < 1.0 then the impact may be considered to have had a negative effect on the bacterial group in question.

Microbial composition in a water sample can be indicated using the RX data. In the BART™ biodetectors, the RX patterns obtained can be used to indicate to some extent the types of microbial species present within the sample. Shifts in the primary (RX1, first reaction signal), secondary (RX2, second reaction system) and tertiary (RX3, third reaction pattern) observed between comparable B and Z sample conditions can indicate that a floristic shift has occurred in the impacted soil. A commonality in the sequence of observed reaction patterns between the Z and B samples would indicate that the perceived impact being investigated has not materially shifted qualitatively the composition of the microflora.

Aerobic or anaerobic dominance in the water may also be determinable. The potential for predominantly oxidative (aerobic) or reductive (anaerobic) conditions has already been addressed with respect to the VVOF shifts which could occur between the B and Z samples. Other signals may also be present which may flag an aerobic or anaerobic dominance. Anaerobic conditions may be occurring where the SRB- *dd* values become shorter in the Z sample; there is a dominance of RX1 of 5 in the IRB biodetector tests; the *dd* for the IRB- tests are shorter or equal to those for the SLYM-; RX2 or RX3 for a SLYM- or an IRB- terminate with an reaction pattern of 6 or 10 respectively. Conversely, a very rapid and short *dd* for the SLYM- with an RX of 1, 2, 3 or 4; an RX in an IRB- of 8 or 9; and absence of any SRB reactions; and a TAB- *dd* of < 3.0 would all indicate that the conditions within the econiches are likely to be aerobic.

MICROBIAL AGGRESSIVITY DETERMINATION (MAD)

Introduction

Some test methods applied to determine the size and activity of the incumbent microflora in a sample of water involve an evaluation of the rate with which they can metabolize and/or grow within a set of defined conditions to generate an observable change or reaction. Present advances would make it possible to detect plasmids of concern in groundwater bacteria[61]. Such observations may involve the simple generation of cloudiness (turbidity) due to cellular growth or changes in the color of the test medium as a result of oxidation, reduction or enzymic metabolic events. The time of delay and the patterns displayed during and after the primary display can be linked to the level of aggressivity of the incumbent microflora which are able to function under the conditions established by the test procedure. Aggressivity does not necessarily relate to the number of viable entities which occur in the water but can be related to the ability with which those entities can adapt to the test conditions imposed. An entity may be considered to range from a single cell, a conglomerate of cells all of the same species type, a consortium of diverse cell types within a fragment of suspended particulate material sheared from a biofilm, ultramicrobacteria and/or microbial spores. The total number of viable cells may be much greater than the number of viable entities. A measurement of the aggressivity allows the observer to appreciate the activity potential of the microflora within the particular water sample.

Traditionally the significance of a particular population of organisms has been directly linked to the number of cellular entities recorded following the assumption that each of these organisms has an equal opportunity to perform. The complex nature of the interactions between the various components within the microflora however distorts the functional ability of each organism to perform to the same extent. It is more probable that the microorganisms with a greater aggressivity for the environmental conditions presented (either in a natural ecosystem or in a laboratory test) would tend to prevail.

One such test system gaining a wider acceptance is the biological activity reaction test system known as the patented BART™ (Droycon Bioconcepts Inc., Regina, Canada). The BART™ systems have been developed for detecting the presence and aggressivity of selected groups of potential nuisance bacteria. The patented test method was originally designed to be a relatively simple presence-absence (P/A) test appropriate to specific bacterial groups. However, in the practical application of these technologies, it was noted that there was sometimes a significant period of delay before positivity (observable event) was obtained (Figure 39). This

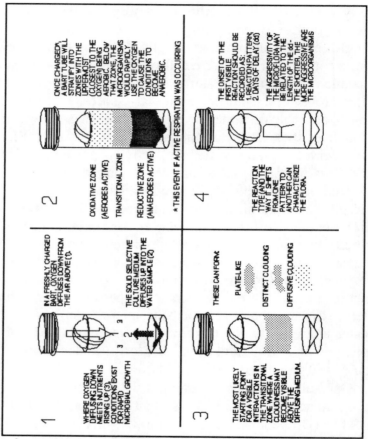

Figure 39. *Generation of a BART™ reaction (sideways, lower left, clockwise). Once a positive tube is charged, it will form an oxidation reduction column (left) in which reactions will occur (right).*

became known as time of delay and generally was measurable as the "days of delay" (*dd*). By experiment, it was found that the *dd* could be used to reflect the size of the original incumbent population for the selected microbial group under test[1,51].

These tests form another simple method (rather than spreadplate population determination) to determine the activity potential of the various groups of bacteria by their potential aggressivity. This may be done using the BART™ biodetectors. Two forms of data can be obtained by this method; they are days of delay (*dd*) and the reaction type (**RX**) which occurs. It can be interpreted that the shorter the *dd* the greater the population size and activity level. At the same time the **RX** obtained gives an indication of the types of bacteria present in the water sample. There are a range of BART™ biodetectors which can be used to examine the presence and aggressivity of selected groups of bacteria in water. Each system when applied in the BAQC-16 format will determine the activity of a different component bacterial group which may be involved in the fouling of the water well. Each system will be considered in turn (see also video presentation disk B).

TAB-BART™: a simple system (total aerobic bacterial ... methylene blue) which will indicate whether aerobic bacterial activity is occurring in the water sample by the rate at which methylene blue is reduced to a colorless form. Where there is a short *dd* (i.e., 1 to 5 days) a significant aggressivity may be occurring. However, where the reduction occurs only after a prolonged time (>5 days) then the aggressivity of the bacteria may be considered to be very low.

IRB-BART™: one of the most complex of the biodetectors in the series since it is able to respond to a range of different bacterial groups which become partially differentiable as a result of the pattern of RX values observed. Particularly of interest is the observation that an RX of 5 may be a reliable indicator of anaerobically dominated biozones and is paralleled by more aggressive (i.e., shorter *dd*) SRB-BART™ reactions. RX values of 8 or 9 may relate to aggressive pseudomonad populations while an RX of 10 (particularly with a *dd* of <3) would indicate a significant enteric bacterial presence (check the GAB:CLF ratio using the spreadplate enumeration technique). A dominance of IRB may often cause an RX reaction of 2, 3 or 4. In cases where the biofouling is dominated by IRB within a particular biozone, the *dd* obtained for the IRB-BART™ will be shorter than that obtained for the SLYM- or FLOR-BART™. Where the heterotrophic bacteria (which do not take up excessive amounts of iron) dominate the reverse scenario would be true. Very often the IRB dominate close to the well while the GAB and SRB dominate further away from the facility.

SRB-BART™: this biodetector detects the sulfate reducing bacteria by the generation of black sulfides which may form precipitates (RX, 1); accumulate under the FID (RX, 2) or throughout the detector (RX, 3). When these biodetectors react in less than 3 *dd*, there is a very aggressive SRB population which is usually found to be associated with even more rapid *dd* in either the FLOR- or the SLYM-BART™. This may be because the SRB commonly cohabit in the anaerobic niches of a consortium which may be dominated by other heterotrophic bacteria.

FLOR-BART™: this biodetector reacts to the strictly aerobic bacteria (such as the pseudomonads) the most rapidly. These bacteria dominate in biozones where there is available oxygen or nitrates, and a restricted range of organic compounds (e.g., gasoline, solvents) in relatively high concentrations (e.g., 5 to 500 ppm). Under these conditions of organically polluted groundwaters, a restricted range of pseudomonads may dominate the biozone and trigger very short *dd* with the FLOR-BART™. Under extreme circumstances, a severely polluted water well has been known to give a *dd* of 20 minutes.

The family of BART™ biodetectors have been designed to allow comparative examinations of potentially fouled water wells. Such *dd* and RX data as is obtained can be used to compare over time whether a biofouling is getting more serious (i.e., a shortening *dd*), responding to the treatment (i.e., a lengthening *dd*) or changing in bacterial composition (i.e., shifting RX and *dd* values). Within the BAQC-16 procedure all of these changes can be used to determine the sequence and composition of the biozones around a potentially biofouled water well. Where necessary the BART™-SOFT program can be used to project the active bacterial population recorded as cfu/mL.

Quantitative Evaluations Using the BART™ Biotechnology

Traditional techniques have involved the examination of water microscopically for the presence of discernible microbial components (such as the iron bacteria), or culturing the viable units on a gel surface containing appropriate nutrients in the culturing medium. This allows a discernible growth of bacteria to form a series of countable colonies. The numbers of microorganisms present in the water is related directly to the number of colony forming units (cfu)[46]. The application of either an experienced eye, the microscopic examination of the particulates, or the utilization of standard technologies requires that these assessment be conducted using trained personnel under standardized conditions.

However, under field conditions, it is not necessarily practicable to employ such standardized techniques for the initial monitoring of the level of microbial activity occurring within a groundwater system and the associated water wells, transmission lines and treatment systems.

The BART™ system offers a technology which can be readily applied by the informed user to determine the activity levels for specific groups of bacteria. These can be separated by the utilization of different BART™ tubes which contain cultural media applicable to the growth and/or metabolic reactivity of the target organisms. The object of applying a BART™ test is to obtain a semi-quantitative understanding of the population size for the targeted microbial group and, by the reactions generated, determine more precisely in a semi- qualitative way the metabolic group of microorganisms dominating actively in the test.

The theoretical application of this patented technology relies on the generation of two zones of microbial activity (aerobic, upper zone; anaerobic, lower zone; parallel: Winogradsky's column). The uppermost zone of activity in a BART™ tube charged with a water sample is at the interfaces between the floating intercedent device (FID commonly in the shape of a sphere) and the wall of the tube. At this site, oxygen from the air above the FID is able to diffuse down into the water between the device and the walls of the tube. This not only encourages the growth of aerobic (oxygen-loving bacteria) within this relatively plentiful supply of oxygen but the surfaces presented by the FID and the wall of the tube present a surface area onto which (sessile) bacteria can attach.

Frequently, these attached bacteria forming a ring around the FID will generate a visible slime or generate a distinctive color reaction. At the same time as these aerobic organisms are utilizing the oxygen around the FID they are also excluding the potential for oxygen to diffuse down into the second strata of the test which is in the main body of water beneath the FID. Any residual oxygen present within this zone would be rapidly utilized by the incumbent aerobic bacterial flora present in the water. Oxygen concentrations therefore rapidly decrease to such a significant level that oxygen is virtually absent from this part of the test medium. When this occurs, the cultural conditions in the lower zone now support the growth of anaerobic microorganisms (these organisms are able to grow in the absence of oxygen and some of them are, in fact, inhibited by the presence of this molecule).

Growth and color reactions are therefore observed in these two stratified zones of the liquid medium. On some occasions, there may be a passive sedimentation of some product of growth (such as dense particulates or chemical deposits like sulfides). Thus, in examining the reactions that can occur in a BART™ there are three zones to examine visually for evidence of microbial activity (i.e., around FID, in main column and deposits on floor).

The speed with which the reaction occurs can be linked to both the activity level and population size of the incumbent microorganisms in the water sample. Where there is a large population or a metabolically very active population, a rapid series of reactions may develop to a level that becomes observable very quickly. The speed with which these reactions can occur may be used to determine approximately the size and activity level for the incumbent bacterial population. In general, the length of delay before the observation of the first reaction is referred to by the number of days of delay (*dd*).

An expression of these activity and reaction patterns in

the test system can therefore be performed and can be recorded using the *dd* data to indicate the relative activity level and size of the population while the reaction type (RX) observed may be further used to semi-qualitatively indicate the dominant types of bacteria causing these activities. There are in all a number of reaction patterns for each of the current BART™ systems. They are for the IRB-, SLYM-, SRB-, FLOR- and the TAB-BART™ systems, a total of nine, six, three, two and one commonly observed reaction patterns, respectively.

To establish a BART™ test on a given water sample, the technology is relatively simple. A water sample is taken and conserved in a sterile container until the BART™ test is performed using the normal standard procedures. When the test is to be initiated, which should happen as quickly as possible after taking the sample, the cap on the inner BART™ tube is removed and water admitted until the FID rises to the predesignated level indicating that 15 mL of water has been admitted to the test vessel. The potential microbial indicators are, at this point, in a state of suspension throughout the sample. It is important **not to shake** the BART™ tube at this time since the culture medium which will cause a selective growth must be allowed to passively diffuse into the water. A radical solublization of the nutrient components constituting the media may create a nutrient "shock" which could inhibit the growth of some of the otherwise active microorganisms and generate a slow or negative result. As soon as the water has been admitted to the test, the cap is screwed down tightly onto the BART™ tube which is, if required, returned to the outer culture chamber which is itself screwed down shut. As soon as the water has been admitted to the BART™ tube and the nutrient chemicals have begun to diffuse, active metabolism and growth is likely to occur. This activity will be affected by the temperature at which the BART™ tube is maintained. The tube may be kept in a safe place away from the risk of being knocked over, or continuously disturbed by persons not directly associated with the testing procedure.

In general, the BART™ can be retained at room temperature away from direct sunlight. This allows bacteria which commonly grow at 18 to 25°C (60 to 75°F) to grow effectively. If the water sample has been taken from a water source significantly colder or warmer than room temperature, consideration should be given to undertaking the BART™ test at lower or higher temperatures more closely akin to that which would have been experienced by the microorganisms in their natural environment. On a daily basis, as is convenient, the BART™ should be examined for any reactions that may have occurred.

When a reaction is noted for the first time and related to one of the standard reaction types, it would be recorded together with the date upon which that reaction occurred. This can then be used as an indicator for the level of activity and population size recorded by the specific reactivity for the microorganisms that were present in that particular water sample. If further samples are taken, any divergence in the *dd*, or RX data would indicate that the bacteria populations in the water are undergoing some changes. If the *dd* value were to go down and the reaction number observed changed; it could be extrapolated that the microbial population and activity had, in fact, increased and changed (in the dominant types of organisms present in the water sample). Conversely, the *dd* value went up, the microbial activity/population levels could be considered to be reduced. The BART™ test method therefore provides a simple method for the convenient observation in a semi-quantitative and semi-qualitative way the microbial populations for the specific group of microorganisms evaluated by the particular BART™ test being performed. These monitoring systems are particularly valuable in determining the relative success of a disinfectant or treatment procedure designed to reduce the particular biofouling events that may have been observed within a water well, groundwater system, treatment plant or distribution system.

It is recommended that confirmation of the data obtained from the BART™ test should be undertaken by an accredited microbiology laboratory using the enumeration procedures outlined in the standard procedures as recommended by the standard manual for the examination of water and waste waters presently in its 18th edition and published jointly by the American Public Health Association and the American Water Works Association.

Since an active bacterial culture may have been generated by the application of a BART™ test, the disposal of the test, once completed, should follow the guidelines laid down by the manufacturer. This involves the utilization of microwaving in a dedicated microwave oven to ensure the pasteurization of the sample. An interpretation of the dd and reaction types for the various BART™s are defined below.

From comparative studies with standard extinction dilution spreadplate techniques and appropriate selective agar media for the microbial groups being tested, the following semi-quantitative interpretations can be made based upon the dd obtained:

< 6 hours or $< 0.25dd$ - Very large $> 10^7$ cfu/mL bacterial population.
- Extremely Aggressive Microbial group.

<26 hrs or $<1.1dd$	- Very significant $>10^6$ cfu/mL bacterial population. - Very Aggressive Microbial group.
>1.1 to $<4.1dd$	- Significant $>10^4$ cfu/mL bacterial population. - Aggressive Microbial group.
>4.1 to $<6.1dd$	- >500 cfu/mL bacterial population. - Potentially Aggressive microbial group.
>6.1 to $<14.1dd$	- <500 cfu/mL bacterial group. - Low Aggressivity
$>14.1dd$	- Microbial population traumatized. - Population size not ascertainable - Aggressivity Potential very low.
negative at 42 days	- No Microbial Aggressivity recorded.

These observations are a generalized interpretation of the data obtained from a range of BART™ test systems. A more complete interpretation is provided in Appendix One where the methodology for using the BART™-SOFT MS-DOS compatible software data interpretation and file storage system is given.

On some occasions, there may be a need to simply establish whether the targeted microbial group is present or absent (presence/absence test, P/A test). From the results of field trials conducted to date there would appear to be an acceptable cut-off time where a significant population of the targeted bacteria would appear to be absent. If the P/A test displays a reaction within six days the target microbial group can be considered to be present in an aggressive form. In these cases, it could be considered that there would be a potential risk of a microbial event occurring, subject to satisfactory environmental conditions. For the field operators, this is probably the most significant interpretation and a control and treatment scenario should be initiated.

Qualitative Evaluation Using the BART™ Biotechnologies

One of the unique and patented features of the BART™ test systems (except the SRB-BART™) is the reliance on the differential establishment of both aerobic and anaerobic regimes around and below, respectively, a floating intercedent device (FID). When the test is charged with the 15 mL water sample, the FID floats up to the surface of the water while the dried selective culture medium slowly diffuses up from the base of the test vessel. At the same time oxygen diffuses in from the air above the FID down into the water interfacing between the FID and the walls of the containing vessel.

Once a tube has been charged with the water sample, do not shake the tube. Such violent agitation will disturb the formation of zones of diffusion which are generated as the culture medium diffuses upwards into the water sample. In some of the BART™ test systems, a series of gradients can be seen to establish up through the water and diminish in intensity as the chemicals diffuse and interact with any microbial components. Very often the leading edge of the diffusing culture medium forms a focus site for enhanced bacterial activity where the oxygen diffusing downward meets the nutrients diffusing upward.

Selectivity is controlled to a large extent by the specific nature of the culture medium employed. Each culture medium, by its chemical composition will stimulate the growth of only some of the incumbent bacteria (targeted organisms) while retarding the growth of others (i.e., selective). The site of any observed growth will therefore tend to be a reflection of the qualitative nature of the various component bacterial species present in the water sample used. For example, where this flora is dominated by strictly aerobic (oxygen-requiring) sessile bacteria, the focus site for growth would initially be at the elevating oxygen-nutrient gradient interface. Later aerobic organisms will grow around the FID at the air/water culture medium interface. Facultative anaerobes may tend to generate activity around the ball as well but will also grow in the anaerobic zone of the water culture medium. These anaerobic (oxygen-free) zones will be created by removing the dissolved oxygen through any microbially based aerobic activities. Growth of strictly anaerobic microorganisms (where oxygen is frequently toxic to their activities) usually occurs beneath the elevating oxygen-nutrient gradient zone and may often be "protected" by encasement within slime-forming aerobic bacteria. In most cases, activity may be most commonly witnessed in the base of the charged and incubated tube.

Reactions in these tests therefore reflect a number of events linked to the presence and maturation of microbial communities within the various oxidative (aerobic) and reductive (anaerobic) parts of the charged BART™ test. Each individual water sample is likely to contain a relatively unique microflora which will interact within the test to generate a range of reactions. These reactions may be summarized into the following major reaction types based upon a series of occurrences such as:

- position and form of visible masses of growth
- generation of microbially generated pigments
- evolution of visible entrapped gas bubbles
- production of colored chemical reaction products such as

    reduced or oxidized forms of iron and metallic sulfides
- odoriferous compounds which may permeate out of the culture vessel
- competition between the components within the microbial consortia present in the water sample

It should be noted that the use of a double-walled vessel in the BART™ test system is essential to reduce the permeation of odors out into the general environment where the tests are being performed. It also reduces the risk of an accidental bleeding of the cultured material out of the test system. The double walls therefore form a containment system for the comfort and safety of the user.

Operating temperatures for conducting these tests should relate to the original temperature of the water. In general, it may be assumed that the bacteria incumbent in the water have either been growing in the waters within the system at the ambient temperatures or are moving through the system in a quiescent mode and not necessarily able to function at the ambient temperatures. Primary interest with respect to biofouling would naturally focus on those microorganisms able to grow at ambient temperatures. Where this is the case then it is important to attempt to conduct the tests under similar temperature regimes. Incorrect incubation (growing) temperatures could distort the data obtained towards non-targeted organisms. In general, for temperate waters operating over temperature ranges from 0 to 30°C, it is often very convenient to utilize room temperatures (18 to 25°C) for testing where there are no incubation facilities. However, where the water is consistently below 15°C there may be a majority of the microorganisms unable to grow at these "warm" room temperatures. In these cases the level of aggressivity recorded may be lower than would have occurred at lower and more appropriate temperatures. In such cases, consideration should be given to incubating the biodetectors at temperatures in the range of 10 to 12°C. Water sampled from sites having much higher ambient temperatures (e.g., whirlpools at 42°C, heat exchangers at 65°C) would require incubation at higher temperatures to more accurately reflect the aggressivity of the microflora being monitored. As a general rule the temperature for incubating a test should be within a range from 5°C below the observed temperature or 10°C above. For warm waters such as heated swimming pools and whirlpools, an appropriate temperature would be 35 to 37°C.

The reactions obtained once positivity has been obtained may range through three major forms. These are: (1) patterns of growth, (2) shifts in color resulting from growths or physico-chemical reactions and (3) generation of detectable gas. Each of these reactions is unique in some senses to each individual test but they can be recorded using the range of standardized test reactions which have been developed for each of the BART™ detectors. While an initial reaction may be noted and recorded, the incumbent microflora may subsequently cause a shift in the reaction patterns as the various viable entities in the community within the biodetector compete for dominance. The pattern of these reaction shifts is a reflection of these community interactions and can be used to aid in the recognition of the floristic composition of that particular flora.

To aid in the recording of these shifts, reaction charts may be used. These were developed based on the experiences achieved in the field and laboratory studies during the research and development of the BART™ technologies[1,18-,22,50,]. In general, the reaction charts are relevant where the incubation temperatures used were at room temperatures (18 to 26°C) as a matter of convenience to the user. It has been found that at higher or lower temperatures there are distinctive shifts in the reaction patterns generated by the same microflora. For example, reaction 5 becomes a more dominant reaction pattern in the IRB-BART™ where 35 to 37°C is used as the incubation temperature.

In practice, the charged BART™ biodetectors can be simply placed on a non-porous shelf and observed daily *in situ*. For some of the tests where a larger and more aggressive microflora is expected (e.g., whirlpools, heat exchangers), more frequent observations may be applicable in the first two days of the test period of incubation.

Interpretation

A positive BART™ is referred to by the days of delay *dd* to the first positive reaction observed and the reaction triggering the event which is recorded from a standard set of reaction patterns known as reaction type numbers or RX. This information can be logged, filed and interpreted using the BART™-SOFT MS-DOS based software. Instructions for the use of this software are included in Appendix Three. For the user, the information so obtained can be used to indicate the level of microbial activity and/or population size (*dd*) as an aggressivity MAD while a characterization of the types of microorganisms by the RX value observed. The RX recorded value may change upon subsequent incubation as other microorganisms now dominate the consortial activities. These shifts in the RX can help to characterize the microbial components in a semi-qualitative manner.

From the laboratory and field experiences to date, a number of characterizations can be generated for specific BART™ tests. Confirmation of these projections should be undertaken using the classical microbiological technologies or submitting a replicate water sample or a reactive BART™ test to a recognized microbiological laboratory able to perform these confirmatory techniques. Identification of the causative organisms can be undertaken by such techniques as the Hewlett Packard G.C./Microbial I.D. (MIDI) system. Information on the floristic composition of some of the BART™ tests targeting bacteria is addressed below.

TAB-BART™

TAB is a dye reduction test method for determining the aggressivity of the total aerobic bacteria. This test emulates the well-recognized methylene blue reductase test widely employed in the dairy industry as a rough estimate of the number of viable organisms in the food[12]. In water, as in foods, aerobic microorganisms obtain energy from the oxidation of organic nutrients which, in turn, creates a more reductive environment. As the habitat becomes more reduced, the methylene blue shifts from the (blue) oxidized form to the (colorless) reduced state (Figure 40). The rate at which this decolorization occurs may be related in a crude way to the activity level of the aerobic bacteria in the water. Methylene blue is positively charged (cationic) which causes the dye to become attracted to, and bond with, negatively charged surfaces such as the bacterial cell wall. In the test for total aerobic bacteria, a series of commonly oxidizable nutrients are presented in the BART™ medium which encourages many of the aerobic bacteria to become active and begin to cause reduction-oxidation potential to shift to a more reduced state. Under these conditions, the methylene blue becomes reduced so that the medium changes from a blue to a colorless form. Since reduction begins where there is a maximum of oxidation stress, the colorless form usually first appears (as a "bleaching" of the medium) at the bottom

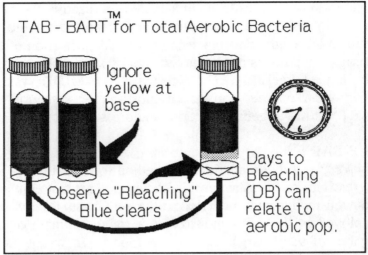

Figure 40. *Methylene blue is reduced to a colorless form by the activities of aerobic bacteria. The bleaching begins at the most reduced site (commonly the bottom of the tube).*

of the test. As the reduction zone extends upwards, so the zone of "bleaching" elevates up through the test culture medium. The most significant event is the time at which the initial start of the "bleaching" process occurs (days to bleaching, DB). A water with a very aggressive aerobic bacterial population will cause a DB event in less than four days. Facultatively anaerobic bacteria (those which can also grow in the absence of oxygen or nitrate) can cause methylene blue reduction but, here, it is reversible and the blue dye color can return to the cultured medium over time (usually at the aerobic interface first). In the TAB-BART™, the culture medium used is very selective for the strictly aerobic bacteria and, in particular, the (section four of the Manual) pseudomonad bacteria. This selectivity reduces the risk of a recovery of the blue coloration associated with facultative anaerobic bacterial activities.

This test can be used to determine the aggressivity of the total aerobic bacteria within a biofouling event. Where the event is dominated by (reductive) anaerobic microbial activities then bleaching may not occur at all or be considerably delayed (DB > 6). In events dominated by (oxidative) aerobic activities then the level of this activity may be crudely observable by the DB which will commonly be < 6. The lower the DB value, the greater the aggressivity of the total aerobic bacteria. In hazardous waste site restorations involving aerobically driven biodegradation, methylene blue dye reduction by the TAB-BART™ may be very rapid due to the high populations of pseudomonad bacteria commonly present (see SHOW TAB).

IRB-BART™ for Iron Related Bacteria (IRB)

The target group for this test are those heterotrophic bacteria able to assimilate, accumulate or in other ways change the forms of inorganic iron presented in the medium (see Figure 41). This may include reduction (anaerobic zone) and oxidation (aerobic zone) sometimes in association with biofilm formations. Note that in this test, the generation of a white or grey cloudiness (turbidity) is not considered significant as an indicator for the presence of IRB bacteria. Iron has to be involved and affect the color of the reaction for the presence of this bacterial group to be confirmed (see SHOW IRB).

Reaction 1 IRB

A dense precipitated slime settled down into the base of the tube, usually mid to dark brown in color. This slime swirls upwards in a coherent ribbon-like manner before dispersing when the tube is gently rotated. Sometimes the

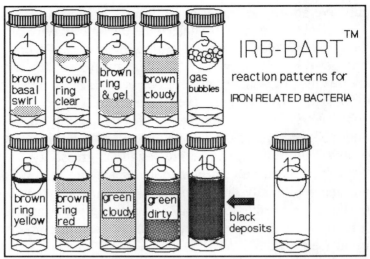

Figure 41. *Eleven reaction patterns are shown above for the iron related bacteria. The presence of iron in the medium causes a variety of color reactions as well as growth forms.*

slime will then break up and disperse but on other occasions the slime will be sufficiently dense to settle down again. This reaction is relatively rare and has been observed in approximately 5% of the cases of a positive reaction to date. The bacteria are probably facultatively anaerobic, able to accumulate oxidized forms of iron into the heavy precipitate, are consortial in nature including section 4, 5, 12 and 13 bacteria. Extracellular polymeric substances (ECPS) would have been produced in quantities adequate to produce a dense slime matrix.

Reaction 2 IRB

The development of a loose slimy brown ring around the FID with no apparent changes in the clarity (usually turning clear) or color of the medium indicates that a sessile strictly aerobic bacterial growth is occurring within which the oxidized iron is accumulating. A mold or actinomycete may also be present. Where this event occurs, the FID may become tightly bound to the wall and form a plug seal (see Universal Reaction 13 described below). This is a relatively rare reaction with an approximately 5% occurrence. In the event of the mold or actinomycete forming a fuzzy cotton-like growth extension over and above the FID, this would be strong evidence of a major fungal or mycelial microbial component in the microflora.

Reaction 3 IRB

A dense slimy matrix forms to occupy the lower 20 to 50% of the water volume in the test. This gelatinous matrix is usually a brown color and retains integrity when the tube is gently rotated or slanted. As the slime does break up there are large flakes of mucoidal particulates flaking upwards into the liquid above the slime. At the same time, a brown ring of slimy growth commonly occurs around the FID. This is not a common primary RX but it is a very common secondary RX. The ECPS being produced in the slime is very water retentive and forms a tight gel. This can incorporate a large percentage of the water into a medium brown gel. When the tube is slanted, the gelled mass will tilt but not necessarily fracture. On rare occasions, the gel mass will elevate and sink within the water phase as the gel density changes. Section 5 bacteria have been linked to this RX. In particular, bacteria belonging to *Enterobacter agglomerans* will generate this reaction but *E. cloacae* generally will not. Ten percent of the tests have been found to generate this reaction as a secondary RX but only 3% have generated this as a primary reaction.

Reaction 4 IRB

The classical reaction first developed in the 1970s was a brown turbid reaction with some pellicular growth at the air/water interface[18]. This has been found to occur in 20% of the reactions and has been found to be a consortial form of growth which can involve section 4, 5, 12, 13, 14, together with sheathed and stalked iron bacteria. *Crenothrix* and *Sphaerotilus* species have been seen in these associations growing in low numbers. Often this RX will shift to a reaction 3 after continued incubation if there are section 5 bacteria beginning to dominate.

Reaction 5 IRB

Very distinctive accumulations of small gas bubbles are generated ranging in size from 1 to 3 mm. These form predominantly around the submerged part of the FID. Each gas bubble may be seen to have an intensely yellow or brownish coating around the bubble which interconnects it to any neighboring bubbles. When the side of the BART™ is tapped lightly with the fingernail, the bubbles will be seen to be relatively rigidly positioned and cannot be dispersed easily. At elevated incubation temperatures, this RX is extremely common (40+%) but at ambient room temperatures, around 10% gave this reaction. Bacteria from sections 5, 6 and 15 may be associated with this event. On some occasions, there has been a correlation between the occurrence of sulfate reducing bacteria in well water discharges and the occurrence of an RX 5 suggesting that this reaction may occur more commonly when the water sample originates from anaerobic econiches.

Reaction 6 IRB

A clear yellow liquid is generated in the nonreactive IRB test (control) and it is sometimes very difficult to distinguish a "sterile" control from IRB RX 6. This reaction is recognized by an obvious but very thin brown iridescent ring occurring at the medium/air interface around the FID or cloudiness in the liquid medium. The ring, while thin (less than 2 mm in depth), can cause a total sealing (occlusion) and will bond the FID to the walls of the tube to form a biological "dam". This bonding may last for one to five days and the BART™ can be physically turned upside down. The liquid medium will not flow past the FID but will remain suspended above the occlusive ring. This seal has been found to withstand an hydraulic pressure created by a 120 mm head of water. This reaction occurs in 5% of the positives thus far and is caused primarily by section 5 bacteria. *Citrobacter freundii* has been frequently isolated as the causative agent for this reaction type.

Reaction 7 IRB

In this form of reaction, the medium darkens to a deep or scarlet red hue during or after which there is often the development of some cloudiness which will darken the reaction. There may be a brown ring at the FID/wall interface. This RX may subsequently shift to an RX of either 3 or 4. Some section 5 enteric bacteria have been isolated from this reaction. The two major genera which can cause this are *Klebsiella* and *Enterobacter*. Where the former genus dominates, reaction 7 may remain stable for a pro-longed period of time (e.g., 25 days). It is a quite common reaction with 10% of the primary RX.

Reaction 8 IRB

The medium generates a clear green coloration which may be, on different occasions, primarily generated at the base and/or around the FID before becoming complete throughout the liquid medium. Over time the green may darken and become turbid, sometimes shifting to RX 9. Section 4 bacteria (pseudomonads) tend to dominate when this reaction occurs. Where there is no subsequent gener-ation of turbidity, *Acinetobacter calcoaceticus* may be found to be a dominant bacterium. Where turbidity is generated with a darkening of the green color the RX value shifts to 9. Species found to be dominant in these events include: *Pseudomonas aureofaciens*, *Ps. putida*, *Ps. syringae*, *Ps. stutzeri*, *Ps. picketii*, *Ps. fluorescens*, *Ps. diminuta* and *Ps. vesicularis*. This common RX has a 30+ % occurrence.

Reaction 9 IRB

A very dark cloudy greenish color which is best viewed by holding the test vessel up to the light. Look through the reaction vessel at a light to see the color. At first glance the test result may appear to be black. This reaction may incorporate either black particles along the walls of the tube or a deep blackened zone at the bottom. This rarely occurs as a primary RX but will often follow a reaction 8. Where it is a primary reaction, there may be a bacterial consortium involved which could include section 5 and/or 7 bacteria as well as section 4. The cause of the continued darkening is possibly due to the generation of bacterial pigments which often occurs during the stationary growth phase. *Serratia marcescens* has been isolated on occasions where this dark cloudy green has occurred. If there is a black pigmentation, probably this is due to metallic sulfides being deposited. This is a rare primary RX (3%) event but quite a common secondary reaction to RX 8.

Reaction 10 IRB

Where there is a generation of a blackened reaction without the occurrence of any green coloration, this is referred to as RX 10. Over time, the liquid phase will become clear and show no distinct color nor even cloudiness on some occasions. It is a more common secondary RX and can indicate the presence of either enteric bacteria or SRBs.

General Interpretation

The IRB-BART™ reactions are thus able to be used to distinguish a number of potential consortial groups of microorganisms. These may be grouped depending upon the RX patterns first observed. The initial generalization would be for the various groupings: 1 to 4 (iron related bacterial dominance); 5 (anaerobic dominance); 6 and 7 (enteric bacterial dominance); 8 and 9 (pseudomonad dominance); and 10 (mixed pseudomonad, enteric flora). The fecal coliform bacterium, *Escherichia coli* (when present in the water sample being tested) will sometimes produce either a weak RX of 4 or 5. The IRB-BART™ has not been designed to specifically respond to the coliforms but rather to the iron related bacteria which have been associated with many water biofouling events. Molds (fungi) may grow in the IRB- BART™ test to form either floating defined

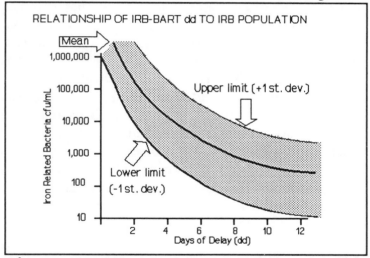

Figure 42. *Relationship between the dd (horizontal axis) for the IRB-BART™ and the population of iron related bacteria (cfu/mL, vertical axis). Range (shaded area) is set at 1 st. dev.*

condensed masses, loosely defined (fuzzy) mycelial sheets or tight woolly rings around the FID (RX 13). They do not usually accumulate the iron and thus will not appear to be of brownish hue. Populations may be estimated semi-quantitatively (Figure 42) or calculated using the BART-SOFT program.

Reaction 13 (Universal)

In situations where the water sample has been collected (Figure 43) from unsaturated porous media or permanently soaked surfaces, there is a greater probability of fungi or mycelial bacteria growing in these environments. When these microorganisms are present in any of the BART™ test systems, there is a potential for these organisms to grow over the FID as either a loose cotton-like structure or a tight "felt pad". These structures may or may not completely cover the FID but they are reasonably easy to observe because of the openness of the structure. Where there is a tighter "felt pad" produced particularly at the FID-wall interface, the growth may become occlusive so that the reaction vessel can be inverted and the FID will be seen to remain biologically "glued" in position. This is a similar effect to the more casual occurrence in RX 6 but will last for a much longer period of time. For example, a *Penicillium* species has been observed to maintain an intact adhesion which could resist a 300 mm hydraulic head for a period of three months. This reaction (13) may occur in **any** of the BART™ series biodetectors.

SLYM-BART™ for slime-forming bacteria

The slime-forming bacterial group is defined as those bacteria which are able to produce ECPS in different forms that will produce some type of cohesive water-retaining growth mass (slime, Figure 43). These bacteria do not necessarily have the ability to accumulate iron in any way other than for basic metabolic function. Many slime-forming bacteria are not iron related in their manner of growth. In practical field studies, there have been many occasions where the biofouling has not been caused by IRB but by the non-iron accumulating slime forming bacteria. This test addresses the needs to monitor this group. ECPS producers are found in sections 4, 5, 6, 7, 12, 13, 14, 15 of the bacterial kingdom and also amongst the gliding, sheath and stalk forming bacteria (see SHOW SLYM).

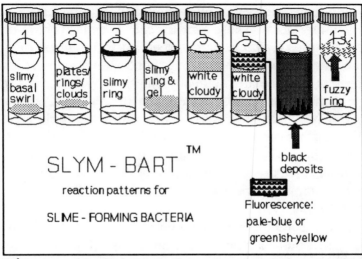

Figure 43. *Slime-forming bacteria commonly do not produce pigments (color). The reactions above are the range usually seen in a SLYM-BART™.*

Reaction 1 SLYM

A slimy growth occurs in the very bottom of the base cone of the reaction vessel. It may not be seen until the tube is gently shaken. It is recommended that all SLYM-BART™ tubes are gently rotated after 24 hours of incubation if there are no obvious signs of growth to check for this occurrence. Very often the slime will spiral upwards in a rotatory manner and then partially disperse to cloud up the medium. This is a very common primary reaction occurring in approximately 40% of the cases.

Reaction 2 SLYM

Vertical or lateral streaks or floating plates of cloudiness may occur where the sessile bacteria are beginning to grow into slime-forming conglomerates. On some occasions, the lateral plates may form at either the diffusing culture medium interface (which rises upwards) or at the descending oxygen interfaces. These growths can form into entire disc-like structures which remain suspended for a period of time in the reactions vessel expanding, contracting and eventually dispersing to generate a cloudy condition. These discs may vary in thickness from 0.5 to 3 mm in thickness. *Proteus vulgaris* has been observed to generate a multiple-layered plating during the earlier phases of growth in the reaction vessel. Illumination from below aids in the observation of this event. This reaction is also very common (30%) and is generally rapidly superseded by a secondary RX.

Reaction 3 SLYM

Here, a light general cloudiness is rapidly followed by the production of a slime ring around the FID which may be very variable in thickness, texture and tenacity. It may or may not be pigmented. Common pigment patterns in the slime ring range through the yellows to pinks. On some occasions, an intense purple color may be generated on the upper face of the slime ring. This may be caused by *Chromobacterium* or *Janthinobacterium* species and may take two to three more days to fully generate after the ring has formed. On some occasions, molds or mycelial bacteria may also grow as an integral part of the slime ring to give a secondary RX 13. Where this happens the texture of the slime will take on a more woolly texture and the pigment, when it occurs, may be of a grey or light blue color. When the molds are involved, there may also be a sealing of the FID-wall interface. When this occurs, the plug will maintain position when the reaction vessel is inverted. These slime rings appear to be produced by consortial associations of bacteria rather than any specific strains. Sections 4, 5, 12 and 13 appear on various occasions sometimes together with yeast, a moderately common RX (10%).

Reaction 4 SLYM

Extremely cloudy with a slime ring growing copiously around the FID together with a heavy slime deposit which swirls up to (frequently) form a twisting ribbon when the vessel is rotated gently. It often occurs after RX 2 or 3 but can, when there is a very high population of compatible bacteria, occur immediately (5% of the cases). The organisms involved appear to be similar to the RX 3 grouping.

Reaction 5 SLYM

A general cloudiness occurs throughout the medium with no intense formation of either a slime ring or deposit. This reaction is particularly common (Figure 44) when the dominant bacteria are pseudomonads (Section 4). Occasionally fluorescence may be observable when exposing the reaction vessel to ultraviolet light. The glowing colors generated may vary from yellowish to greenish hues or, more rarely, a pale iridescent blue. These pigments occur two to three days after the onset of visible growth, will occur around and just below (5 mm) the FID, and will commonly last for a period of one to six days. Comparative field trials with the standard media for culturing and recognizing fluorescing bacteria (mainly pseudomonad types) have revealed that the fluorescence is generated in only 70% of the cases where the phenomenon was observed on PAF spreadplates using the same water samples. The test seems to be more supportive for the *Pseudomonas fluorescens* group. Approximately 12% of the tests exhibit this RX in field trials. The FLOR- BART™ has been developed to improve the rate of recovery of fluorescent pseudomonads, including *P. aeruginosa*. It should be noted that where *P. aeruginosa* is suspected, there may be a health risk posed and subsequent confirmatory testing should be undertaken following the standard procedures.

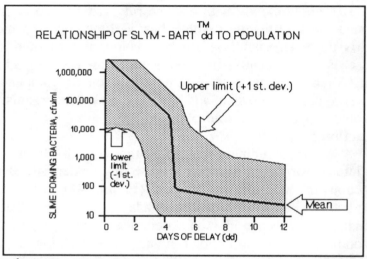

Figure 44. *Relationship between the dd (horizontal axis) for the SLYM-BART™ and the population of slime forming bacteria (cfu/ml, vertical axis). Range (shaded) is set at 1 st. dev.*

Reaction 6 SLYM

The medium rapidly develops a dark green, almost black color with considerable turbidity. There is an ongoing darkening for a period of time before the reaction stabilizes.

This may take a period of five to fifteen days. This is a relatively rare initial test RX pattern (3%) but can occur very commonly as a secondary reaction. When it occurs, the dominant bacterial components appear to be the various pigmented bacteria of sections 4 and 5. It is also possible that a rampant proteolysis may occur which would cause some hydrogen sulfide production and hence metallic sulfide deposition. Parallel tests using an IRB-BART™ may give an RX10 for the same water sample.

SRB-BART™ for the sulfate-reducing bacteria

The sulfate-reducing bacteria are primarily identified by their ability to produce hydrogen sulfide from sulfates. The marker for this event is the deposition of black iron sulfides. This test triggers as positive when such deposition occurs. Each of the RX values are determined by the position and extent of the deposition (Figure 45). Turbidity generation (cloudiness) is not considered to be an integral part of this test procedure (see SHOW SRB).

Reaction 1 SRB

Black sulfide deposition occurs only in the conical base of the tube. This black deposit may be a tight coherent mass or consist of black granules adhering to the sloping sides and lower parts of the walls of the reaction vessel. This is the most common reaction (70%) and a light cloudiness may be noted as the deposit first forms. It generally gets much more intense as the time progresses.

Reaction 2 SRB

An irregular black ring occurs around the lower part of the FID and over to the wall. These deposits consist of ribbon-like, often slimy processes which become strung around the FID and may or may not interconnect. Some nonpigmented slime formation or cloudiness may or may not

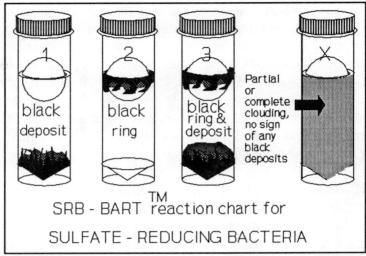

Figure 45. *Sulfate-reducing bacteria are recognized by their ability to generate black iron sulfides which may accumulate in the base (1), at the top (2) or at both sites (3).*

accompany this event which is a much less common reaction than RX 1 (10%).

Reaction 3 SRB

Intense black ring forms around the FID while, at the same time, heavy black deposits form in and above the base cone of the reaction vessel. The medium is usually turbid (cloudy) but with blackened particles often attached to the walls of the vessel or floating freely in the medium. It occurs as the primary RX in 20% of the cases of positivity but is very often a secondary RX for both reactions 1 or 2.

The culture medium used is a general purpose selective medium for the culturing of the SRB. Because the conditions have to be anaerobic to avoid the presence of oxygen which is toxic to these bacteria, the entrance of oxygen is denied by the application of an oxygen-impermeable oil barrier at the FID-wall interface (Figure 46). Occasionally, a slime ring will form around the ball and penetrate this barrier to initiate limited aerobic activity at that site. This test has been designed for the examination for the presence of most of the section 7 bacteria.

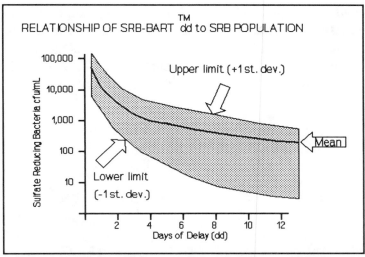

Figure 46. *Relationship between the dd (horizontal axis) for the SRB-BART^TM with the population of sulfate reducing bacteria (cfu/mL, vertical axis). Range (shaded) is set at 1 st. dev.*

Reaction X SRB

Sometimes an intense clouding occurs without any blackening of the medium. While these bacteria are not SRB, they are capable of growing under anaerobic conditions. Such a presence (particularly where other BART^TM biodetectors remain negative) would indicate the presence of anaerobic bacteria.

FLOR-BART™ for the fluorescing pseudomonads

This test incorporates a selective culture medium which specifically encourages the growth and fluorescing activity of species belonging to the genus *Pseudomonas*. There are consequently reactions which relate specifically to these events and are recorded as two major reaction patterns. Growth in this test is reflected in a diverse range of growth patterns which can generate either a cloudiness or a pattern of slime formation (Figure 47). Fluorescence is the transient generation of fluorescing pigments (in U.V. light) which occurs either before (rare) or after (common) the initiation of growth.

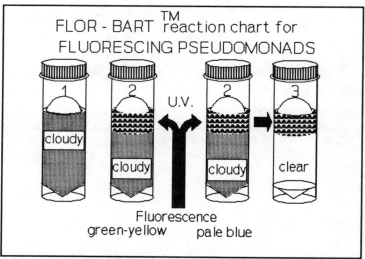

Figure 47. *Pseudomonads are aerobic bacteria which cause a range of fouling problems related to corrosion and plugging (1) and hygiene risk concerns (2 and 3) such as ear, eye and skin infections.*

The biodetection system focuses on the fluorescent

pseudomonads of both major groups (i.e., *P. aeruginosa* and *P. fluorescens*) and the section 4 bacteria likely to be associated with biofouling (see SHOW FLOR).

Reaction 1 FLOR

This reaction occurs when there has been a distinct microbial growth occurring within the reaction vessel. In most cases this growth may be a simple clouding but more complex growth forms can occur. These forms include plating (discs of intense turbidity), threads (spindly webbings slimy in nature extending through the still clear culture medium), gels (dense water-retaining slimes which may occupy a significant volume of the medium), rings (slime growths around the FID)and foams (entrapped gas bubbles). These growths may be pigmented in the yellow, pink or beige color range but without fluorescence.

Reaction 2 FLOR

Fluorescent pigments occurring which are easily observable by exposure to ultraviolet light (follow manufacturer's safety guidelines) and are accompanied by visible evidences of growth as described for reaction RX 1. Normally, these pigments would appear one to three days **after** the initial growth had been observed and would be transient in nature. Normally these pigments may be observed for a period from one to ten days with an average presence of three days. RX will normally follow by one to two days an RX of 1 where fluorescent pseudomonads are present.

Reaction 3 FLOR

On rare occasions there will be the generation of fluorescent pigments around the FID even though there is no visible evidence of microbial growth. Such events would indicate that a very aggressive fluorescing pseudomonad was present within the reaction vessel and was rapidly generating observable levels of fluorescent pigments.

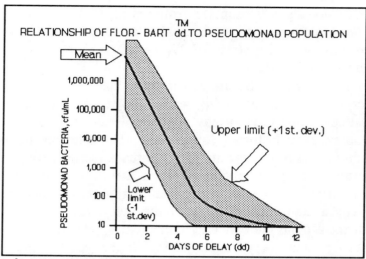

Figure 48. *Relationship between the dd (horizontal axis) for the FLOR-BART™ with the population of pseudomonad bacteria (cfu/mL, vertical axis). Range (shaded) is set at 1 st. dev.*

IN-WELL INCUBATION

The radical changes that occur between the biofouled zone within the groundwater system and the water being sampled downstream from the wellhead can be radically different in characteristics (e.g., temperature stability and range, oxygen concentration, hydraulic flow patterns, redox)[22]. These differences naturally bias the type of microflora which will grow and dominate the water systems. Many researchers have examined the potential to culture the microorganisms actually within the water column of the well itself[49,70,73]. Environmental conditions at such a site are generally considered to be closer to those that would be experienced right at the site of the biofouling (e.g., at the redox fringe between oxidative and reductive conditions back in the gravel pack or natural porous media formations). Two approaches have been proposed to monitor such down-hole activities. In the first approach, the major recognition of microbial presences is through a visible attachment and subsequent growth on some form of immersed coupon (immersed attachment, IA)[72]. This technique relies in essence on the microbial entities physically attaching to the coupon from a suspended state in the water. For the second approach, a volume of water is entrapped within an closed system where the environmental conditions are modified to encourage the selective growth of targeted microorganisms (in-well incubation device, IWID).

Immersed Attachment

IA techniques generally involve the suspension of glass microscope slides in the water column at a midpoint position relative to the screened area of the well. A number of approaches have been proposed which can be subdivided into two groups based on a passive format where no additional chemical stimulation is provided, and an active format where some chemical supplementation is provided.

In the passive format, the most common technique is the direct suspension of glass microscopic slides down the well for a period of 14 days before retrieval[66,72]. Once removed, the slides may be examined microscopically either directly or after staining. Generally, one side of the slide would be selected for observation and the other side would be physically cleaned using paper tissues. If no microbial evidence was observed, immersion periods for these slides should be extended to 28 or even 42 days. Two presentation formats for the slides have been tested. In the Hanert IA technique developed in 1981, three microscopic slides are presented in a vertical array equidistant along a one meter PVC rod with an O.D. of 10 mm. The slides are glued in position with an acrylic filler adhesive so that only one side is exposed outwards for microbial attachment and growth. Since then a number of slide immersion systems have been described[21,22,51,77,72,73]. Generally, these involve the insertion of the slides for 10 to 14 days before recovery and examination.

In the active IA format, some chemical stimulant is provided to encourage attachment and growth. Smith described a modification of the Grainge and Lund procedure which would be applicable down hole[72]. In the original procedure, a simple test system was described[37] using a mild steel non-galvanized washer and a plastic rod to stimulate the appearance of IRB. In the modified IA procedure, a 100 mm PVC rod was used onto which a steel washer was used as both ballast and a source of iron for the IRB. These were suspended down the water column of the well at intervals of 30 cms. Here, the presence of "mobile" iron (i.e., the washer) next to a pristine surface (i.e., the plastic rod) was considered adequate to stimulate growth. Unlike the glass slide techniques, this technique relies upon the occurrences of a visible growth to indicate the presence of IRB. This can then be removed for microscopic examination and culturing as appropriate.

These techniques rely upon the suspending of the devices down in the water column of the well. This on occasion can cause inconvenience to the well operator during the times of insertion and recovery of the coupons. Additionally for the technician, there is the risk of getting "hang up" problems as the devices are lowered or raised from the designated sampling sites. Smith noted that microscopic examination of these growths revealed heavy overburdens of iron which could often make the identification of the IRB difficult. Preference is often given by the operators to position other slides in the wellhead using a modified slide support system. This is more convenient for the recovery of the slides.

In-Well Incubation Device, IWID

While the IA systems, at best, provide a mild stimulation of growth in the IRB community, the IWID system is designed to provide a series of separate down-hole culture vessels. Each of these culture vessels can be charged once in position in the water column of the well and left there for the period of incubation. After the incubation, the whole unit is returned to the surface for a determination of the microbial activity which may have occurred.

The first IWID was developed from 1983 to 1987 at the University of Regina and used to determine the rates of biodegradation of 2,4-D and phenol by the "normal" microflora. In configuration, the device consists of a 1.2 m tube with an I.D. of 100 mm. Arrayed inside the tube are pairs of 15 mL glass vials charged with 1.0 mL of concentrated (x10) culture medium (Figure 49). This medium will dilute to the correct strength once the vial is charged with water (10 mL) from the formation. Inside the vial is a floating hollow plastic ball (9.75 mm diameter) which rises to the surface as the water charges into the vial. The ball cannot float out of the vial because access is restricted by the cap which has a 7 mm diameter hole drilled through the center. This hole allows the water to enter the vial until the rising water level forces the ball up against the hole and prevents more water entering into the vial as well as the culture medium "bleeding" out.

To charge the IWID, the tube is capped at both ends with an air line passing down through the upper cap (AL

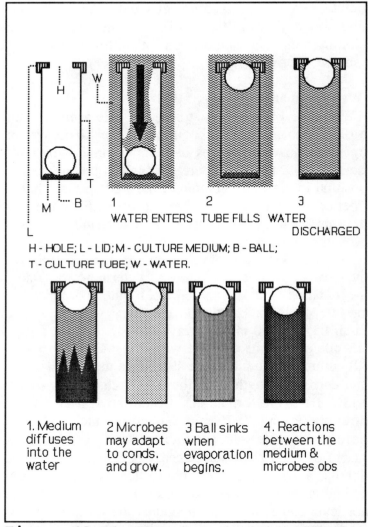

Figure 49. *Schematic mechanism by which an IWID culture tube fills with water (upper row) and then is incubated down hole (lower row) during which reactions and growth may be observed.*

port) while the lower cap is connected to the outside through a sampling port (SP) and carries a 24 mm (I.D.) 300 mm long extension tube which is weighted with 5 Kg metal ballast (to keep the IWID vertical during descent and incubation, Figure 50). The lower exit port on the SP port is sealed with a screw cap (SC) to maintain the sterility of the inside contents of the IWID once it has been prepared for insertion.

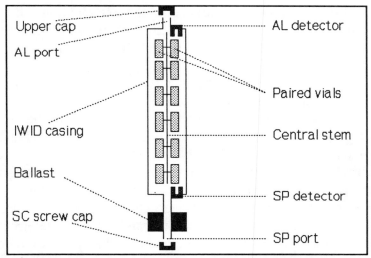

Figure 50. *Vertical section of an In-Well Incubation Device (IWID) supporting six pairs of vials (IWID culture tubes). Conductivity detectors (AL and SP) detect the presence of water.*

Two conductivity detectors are installed inside the uppermost (by the AL port) and at the base inside the tube (just inside of the SP). These detectors record the presence or absence of free water by shifts in the conductivity from high to low respectively when water is present. Where the lower (SP) conductivity meter registers a high conductivity there is free water in the base of the incubator tube and, where the upper (AL) also registers a high conductivity the tube may be considered to be full of water. These conductivity detectors therefore indicate when the incubator tube is full (SP +, AL +), empty (SP -, AL -) or water has leaked into the system (SP +, AL -).

The IWID is made ready for insertion by placing the glass vials into the clip holders around the central insertion stem. The precise selection of the contents of the vials would depend upon the purpose of the examination. If the function was to investigate whether there was an active microbial presence in the water well, a broad spectrum of different cultural formulations could be used. Where there was a need to determine the spectrum and tolerances of the microbial entities involved in a biodegradation function, the selection would be very different. It would already be clear to a person familiar with the art as to whether the degradation was occurring aerobically, anaerobically or by nitrate respiration. Additionally, the target compound(s) of degradation would be known. By including vials containing a range of concentrations of the targeted chemical(s), subsequent degradation patterns obtained can be used to determine the range of optimal degradation and also the upper tolerance concentration limit for this activity. Higher concentrations may be degradable by an adjustment to the nutrient regime, or these concentrations may be lethal to the biodegraders. Subsequent IWID-based investigations can determine these relationships.

The appropriate gas (or air) used to charge the IWID will influence whether an aerobic (using air), anaerobic (using a gas such as nitrogen) or a nitrate respiration (gas plus the addition of at least 30 mgN/L as nitrate to each of the vials) will occur. The gas supply is connected to the AL port and passed through a cotton wool packing (CWP) to filter out any viable particles from the gas or air. A gas vent (AV) valve is installed along the gas line close to the cylinder or compressor. This valve is used to vent gas pressure at the time of charging the IWID with water.

The IWID is charged by lowering the device SP end first down the well while maintaining a positive gas pressure (Figure 51). This causes the gas flow (GF) out of the SP port which prevents water (W) entering. When the desired

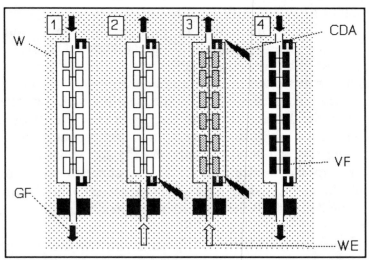

Figure 51. *Schematic of IWID charging with water (W). Gas flow (GF, down arrow) out of the SP port keeps water out except when gas pressure reduced (up arrow) when IWID is charged.*

depth is reached, the gas pressure is reduced, causing water to enter (WE) the IWID and trigger both the lower and upper conductivity probes in turn. Once the upper probe indicates the presence of water, all of the vial (IWID culture tubes would have become charged with water. Gas pressure is now increased, displacing the free water in the IWID with the desired gas and the vials are filled (VF) and incubation can begin either at-site or after removal from the well.

Sterility is a major concern in the assembly of an IWID for down-hole investigations. If there needs to be an absolute assurance that the organisms being monitored originated in the groundwater and not through a "chance" contamination via the IWID, vials or the gas (air) supply, then the IWID and the vials have to be prepared in an aseptic state (i.e., free from contamination by any viable entities) or subsequently sterilized after assembly.

Insertion of the IWID down into the water column involves a number of stages in the preparation. The gas (air) supply has to be connected to the upstream side of the cotton packing so as to sterilize by filtration the gas stream entering the IWID. There should be an adequate pressure generated either by the decompression of the gas in the cylinder or via a compressor to balance any hydrostatic pressures at the sampling site and be able to prevent the accidental entry of water into the device. Both of the conductivity meters should be connected up to the surface meters and checked. Both the AL and SP should read a low conductivity (i.e., no free water in the IWID).

Once positioned over the water column in the well and ready for insertion, the screw cap SP is removed from the lower SP port and the gas (air) flow through the IWID commenced immediately. The IWID is now lowered down the well into the water column and held for one minute as it becomes totally submerged. At this time, the AL and SP conductivity readings are checked and should read low. If the SP is reading a significant conductivity (but not the AL) then water may be leaking slowly into the IWID and the unit should be withdrawn and the problem rectified. If both the SP and the AL exhibit a high conductivity, the leakage is very serious and the IWID would not be operable. Where the SP and AL both show a low conductivity, insertion can continue down to the depth for sampling and incubation (e.g., midpoint of the screened section of the well). A sufficient flow of gas should be maintained to keep water from entering the IWID (i.e., the SP reads negative) but not so much flow that the water in the well appears to "boil" due to all of the gas being vented from the IWID. Gas flow should be just sufficient to maintain a slow trickle of bubbles to the surface from the IWID. As the IWID descends the water column, clearly the gas pressure would have to be increased steadily to compensate for the increasing hydro-static pressures (e.g., if the SP detectors shows an increasing conductivity, raise the gas pressure slightly until the conductivity drops).

Once the IWID is at the predetermined depth, the sampling may now be initiated. This is a phased operation in which the IWID is flooded with water from the column (to charge the vials) and then expelled (to prevent cross contamination between the various vials). It is important to follow the steps carefully to avoid a faulty sampling.

1. Lower the gas pressure slowly by opening the gas vent GV valve until water now begins to trickle into the IWID and the SP registers a high conductivity.
2. Increase the pressure just enough to cause the SP to fall to a moderate conductivity. Water has now entered the IWID and been expelled again. Lower the gas pressure again and note whether water enters the IWID.
3. Time the moment that the water now enters the IWID and continue to gradually lower the pressure so that water now continues to enter and flood the IWID. As the water rises up past each row of vials, water will flood into the vial causing the ball to rise and block any further entry of water.

4. Record the time when water reaches the upper port (i.e., the AL port records a high conductivity). All of the vials would now have become charged with water and become sealed by the floating balls.

5. Increase the gas pressure to force the remaining water out of the IWID and time the period between the AL and the SP ports again registering a low conductivity (water expelled).

6. Adjust the gas flow to just prevent entry of water into the IWID. All of the vials should now be charged and the process of at-site incubation has commenced.

To economize on the amount of gas (from a cylinder) or compressor time being used during incubation, the gas flow can be shut off during the incubation period. Daily inspections are, however, recommended to ensure that the dissolving of the gases in the water is not causing water to enter the IWID (i.e., the SP should continue to register a low conductivity). Where a high conductivity is seen the gas pressure is increased until the SP again shows low conductivity. In an event where the AL port shows a high conductivity, the study has to be abandoned due to probable interconnections between the vials.

Conductivity is addressed in only general terms since this would have to be related to that of the water in the well. A high conductivity would be that of the well water while a low conductivity would that associated with the transient water film interconnecting the probes.

The time delay when filling the IWID can be used to give an indication of how turbulent the water would have been when ascending the inside of the IWID and charging the vials. Too fast a filling may cause excessive turbulence which in turn would increase the likelihood of losses of nutrients from individual vials. This rate of elevation (RE, mm/sec, rate of filling) may be calculated:

$$T_f \quad = \quad T_{al} \quad - \quad T_{sp}$$

where T_f is the length of flow period (seconds) ascending the IWID, T_{sp} is the time at which the SP registered high conductivity and T_{al} the time at which the AL port showed a high conductivity.

The distance between the SP and the AL probes is known as D_{iwid} and is measured in mm. The RE value can now be calculated:

$$RE \quad = \quad T_f \quad / \quad D_{iwid}$$

An ideal RE value would be between 3 and 7 mm/sec with an acceptable range of between 2 and 10 mm/sec. A high RE value would indicate that an increased risk of turbulence would exist.

Incubation periods for the IWID once-charged and static down hole may vary depending upon the nature of the investigation and the water temperature. As a guide to the length of an incubation period, the following relationships may be used where aerobic biodegradation is the subject of the investigation (temperature range C°, incubation length in days): < 12°C, 28 days; 13 to 20°C, 21 days; 21 to 26°C, 14 days; 27 to 34°C, 7 days; 35 to 40°C, 4 days; 41 to 46°C, 10 days; > 47°C, 14 days. Where the investigation is of a possible natural aerobic biofouling event, the length of the incubation period recommended above should be doubled (i.e., x2). For anaerobic biodegradation or biofouling, the length of the incubation period should be tripled (i.e., x3).

At the end of the incubation period, there are a number of stages which have to be followed to retrieve the IWID.

1. Check that both the SP and AL ports are still displaying a

low conductivity.

2. Increase the gas pressure until a gentle stream of bubbles is rising to the surface of the water column.

3. Gently raise (1 to 3 meters/min) the IWID up out of the water column. Note that where anaerobic or nitrate respiration conditions have been employed for the incubation, increase the gas pressure marginally once the IWID leaves the water. This is to prevent oxygen from the air entering the IWID and causing aerobic conditions to be generated.

4. Seal the IWID by closing off the SP port with the SC cap and shutting off the gas supply. The gas line is now disconnected above the CWP filter after the CWP valve has been closed.

Examination of the IWID may be conducted at-site where the objective is to determine whether visible microbial activities occurred in any of the vials. If the IWID has to be examined in a laboratory setting, it would have to be shipped while being kept: (1) in a vertical position; (2) maintained in the same gas composition as was used down-hole for the incubation period; and (3) the relocation time should be minimized and the temperature maintained within 5°C of the down-hole water temperature. If a prolonged time (e.g., > 12 hours) is required to relocate the IWID in a laboratory to be opened, the IWID can be cooled down to 4 to 8°C to minimize any ongoing microbial activity. It must be borne in mind that this "cold shock" may not only retard microbial activities within the individual vials but may also traumatize the incumbent microflora so that prolonged recovery times may be required to recover and culture these organisms.

If anaerobic conditions have been in the IWID, it should be borne in mind that many anaerobic microorganisms are oxygen-sensitive. If the IWID is opened in a laboratory and the vials exposed to oxygen, many of these obligate anaerobes may enter a severe stress condition or be killed. Consequently, where anaerobic conditions have been employed and it is desired to culture the microflora, the IWID should be opened in an anaerobic chamber where oxygen has been excluded.

10

Glossary

Acinetobacter, a gRAM-negative rod-like bacterium which becomes coccoid as the cells age. These bacteria are commonly found in surface waters, tailings ponds and distribution systems and are sometimes called the "weed" of aquatic bacteria. It has been associated with outbreaks of gastroenteritis and linked with waterborne (nosocomial) infections.

Acid-fast, a staining procedure which differentiates out the bacterial genus *Mycobacterium* due to its ability to retain a dye under acid-alcohol washing.

Acidophile, microorganisms which grow most efficiently within the pH range from 1.0 to 5.5.

Actinomycete, aerobic gRAM positive bacterium which grows forming branching filaments (hyphae) and produces exospores.

Aerobe, an organism that grows in the presence of oxygen. Growth may occur only in the presence of oxygen (obligate) or may also occur in the absence of oxygen (facultative).

Aerotolerant, condition where an organism can grow equally well whether oxygen is present or not.

Active bacteria, bacteria that are in a state of either growing and/or metabolizing actively. This occurs when the

327

environmental conditions are favorable for the requirements of the particular group or groups of bacteria involved.

Adaptation, the intrinsic ability of microorganisms to adapt successfully to changes in the environment. Many microorganisms have the facility to adapt their metabolic processes to meet the needs dictated by changing environmental conditions (i.e. nutrient source change).

Aeromonas, a gRAM-negative facultatively anaerobic bacterium related to the coliform bacteria (but not forming a part of that family). Widely associated with diseases of a variety of fish and reptiles. These bacteria can also be present in groundwater systems, usually in very low numbers. The species, *A. hydrophila* has been associated with outbreaks of gastroenteritis and is considered to be a nosocomial pathogen.

Algae, small single or simple multicellular plant-like organisms which grow normally in the presence of light by photosynthesis. Occur widely in fresh water and are related in population size to the degree of nutrients available to the water body. Some are capable of using organic materials for growth (heterotrophy) and may occur in wide bore or shallow wells and in deeper wells where there is an adequate organic nutrient base.

Anaerobes, organisms that can grow in the absence of oxygen. Many of these bacteria are known to be so sensitive to oxygen that the presence of even low levels (i.e., 0.1 mg O_2 /L) will kill the cells.

Antagonistic effects, the effect that one organism can have to the detriment of the survival and/or growth of another organism through the production of a toxic product such as an organic acid or the production of a target-specific toxic agent such as an antibiotic.

Archaebacteria, a group of bacteria which evolved very differently from other bacteria and are found today growing under very unusual environmental conditions (e.g., high saline, high temperatures, methane production zones, and sulfur-rich conditions).

Attached bacteria, those bacteria which are able to attach directly to a surface where reproduction and growth can occur while the cell(s) remain attached to a surface. Usually these organisms grow within a biofilm.

Attachment, the process by which a microorganism can fix itself to a surface usually by the excretion of extracellular polymeric substances (ECPS) such as polysaccharides.

Bacillus, a rod-shaped bacterium.

Bacterial reactants, bacteria which are able to either set up a physical and/or chemical reaction within an environmental system to cause a reaction within that environment which will reduce or exclude the growth of competing organisms.

Bacteriophage, a virus that infects bacteria, sometimes referred to as a phage.

Bacteriostatic, inhibits the growth and reproduction of bacteria.

Bioamplifier, an organism which accelerates (i.e., amplifies) a natural physical and/or chemical reaction causing that reaction to occur at an accelerated rate.

Biocides, specific compounds or groups of compounds which can in either a gaseous, dry, suspended or dissolved state cause a toxic (lethal) effect on a broad spectrum of target organisms.

Biocolloid, a suspended organic particle in which living microorganisms may be surviving and/or growing. Size usually ranges between eight and one hundred microns. Density similar to surrounding aqueous environment.

Bioconversion, the use of living organisms to modify substances to forms which are not subsequently (or normally) used for growth. These product forms may be recoverable. Also known as biotransformation.

Biodegradation enhancement, a modification of the

environmental conditions by such techniques as nutrient enhancement or inoculation of microorganisms, to enhance the degradation of particular compounds of concern.

Biodetector, a device which allows the detection of specific or general presences relatable to various microbial activities.

Biodeterioration, the biologically mediated deterioration or destruction of materials such as paper products, woods, concretes, paints and metals.

Biofilm, a continuous or discontinuous film of microbial growth occurring over a surface in which the individual cells are bound within a common matrix of extracellular polymeric materials. The biofilm may take the form of a thin "greasy" covering, a thick slime or a complete plugging of the interstitial matrix within a given environment.

Biofilm dissolver, an organic or inorganic chemical or product of microbial activities which causes a biofilm to disintegrate and slough off into the associated liquid and/or other phases.

Biofilm-forming microorganisms, that group of microbes which are able to either independently or in a consortium form a continuous biofilm of either a temporary or permanent nature over a given solid surface at the interface with a water-based liquid matrix.

Biofouling, the phenomenon of deteriorations of any type resulting from biological activities (e.g., microbially induced fouling, MIF). For example, the production of a thick biofilm leading to a tubercle with associated corrosion on an iron surface.

Bioincumbency, the density of viable microbial entities within a biofilm or a biocolloid. Usually estimated in viable units/$10^{-12}m^3$.

Biomagnification, the accumulation of a substance within a living system beyond the basic requirement for the incumbent organisms. Sometimes referred to as bioaccumulation.

Biomass, the total mass of biological material (including any attached extracellular organic materials) that are present within a given system. In general, the biomass is estimated as a dry weight which normally forms 10-20% of the wet weight of that given biomass. Carbon normally forms 50% of the total dry weight mass.

Biomass accumulation, the increase in biological growth (both in terms of cellular and extracellular product mass) as a result of growth. This biomass growth may occur in three forms: the planktonic (dispersed), sessile suspended (particulate) and sessile (attached) phases.

Biosensor, the manipulation of a biological process to generate a signal (usually electrical) in the presence of particular substances (such as toxic agents).

Biozone, a localized formation of biological activity involving a specific group of organisms within a community structure. May interface with other biozones.

Bleach, common name for hypochlorite based disinfectants widely used today for the control of both planktonic and sessile bacterial populations. The most common forms used are sodium and calcium hypochlorite and are generally considered to be particularly effective against the coliform bacteria.

Blended chemical heat treatment (BCHT), a patented process in which a severely biofouled water well is recovered by a triple-phased treatment involving the use of heat, disinfectants and acidization to shock, disrupt and disperse the biofouling.

Blended techniques, the utilization of more than one technique to control a given situation wherein the sum of the two control strategies, when applied simultaneously, is greater than that which would be achieved by using both techniques independently.

Campylobacter, gRAM-negative, curved rod-like bacterium considered to be one of the nosocomial bacteria which does occur on occasion in waters.

Clogging, the total restriction in flow of groundwater through an aquifer and on into a water well. This may be the result of the accumulation of such materials as chemical precipitates, silts, clays and extensive biofilm slimes (see also plugging).

Coccus, a bacterial cell which may be spherical or close to spherical in shape.

Coliform, gRAM-negative nonsporing bacteria which are facultative anaerobes and able to ferment lactose to acid and gas within 48 hours at 35°C. Considered to be an excellent biomarker for fecal contamination due to the abundance with which these coliforms occur in fecal material while being less common in the natural environment.

Colony, a population of cells which grow as a cluster or assemblage on a solid surface sufficiently to become visible. Common surfaces used are agar gels or membrane filters.

Colony-forming units (cfu), the enumeration of the number of colonies formed on a given solid surface from a known volume or mass source. It is generally assumed that each colony unit formed represents one viable cell (i.e., each viable cell will form a separate and recordable colony).

Community structure, the form within which a consortium of microorganisms develop into a community structure in terms of the relative layering (stratified) positions of the bacteria one to another. It may also reference the various strains within a community functioning interdependently.

Consortium, a group of organisms which work co-operatively to maintain a common econiche within which all can survive and flourish.

Continuous flow slide culture, the cultivation of an attached biofilm on a surface under a continuously flowing liquid medium. These biofilm cultures can be used to determine the mechanisms of colonization of specific surfaces and subsequently interpret the dynamics of that growth.

Control strategies, any strategy which has been designed and is implemented in order to suppress a given biological problem without the expectation that the problem will be permanently resolved.

Corrosion, the deterioration of solid or porous materials and their surfaces by physical and/or chemical erosive mechanisms which can be catalyzed by the presence of biological amplifiers. For example, the production of hydrogen sulfide by sulfate-reducing bacteria can initiate electrolytic corrosion often accompanied with tuberculation and the eventual disintegration of the structure afflicted (microbially induced corrosion, MIC).

Culture, the observable growth of specific microorganisms in a particular laboratory medium. Usually restricted to a single strain or kind of organism.

Cyanobacteria, bacterial group of photosynthesizing organisms in which oxygen is evolved. Used to be known as the blue-green algae.

Decomposer, an organism that breaks down complex materials into simpler components, including the releases of simple inorganic products.

Denitrification, the reduction of nitrate to nitrite (partial) and to nitrogen gas (complete) during anaerobic metabolic activities. Can cause nitrogen (as a nutrient) to become lost to the ecosystem.

Detached bacteria, any individual or group of bacteria which occurs within the liquid phase and is detached from any solid surface. Commonly able to move either through motility or be carried along with the liquid flow, possibly in biocolloids. These are sometimes referred to as planktonic bacteria.

Diagnostic, the act of determining the cause of a specific phenomenon. The methods may be chemical, physical and/or microbiological.

Dinitrogen fixation, the microbial act of fixating gaseous nitrogen (N_2) either aerobically or anaerobically into a form assimilable by the organisms.

Discharge, water being pumped directly from the water well or seeping out from a spring or relief well.

Disinfection, the use of a physical and/or chemical process for the reduction in the total loading of microorganisms and the elimination of any (perceived nuisance) organisms thought to be associated with a waterborne concern.

Endospore, a condensed form of an individual bacterial cell which is extremely resistant to heat and other harmful agents. It is a mechanism **for survival and not reproduction.**

E.coli, abbreviated form of *Escherichia coli* (esh"er-i'ke-ah ko'li), a gRAM-negative facultatively anaerobic enteric bacterium which forms one of the dominant bacteria in human feces and is the most frequently used marker organism for the indication of fecal pollution of food, surfaces and waters. Sometimes referred to as fecal coli (FC).

Encrustation, occurs as a thick irregular surface coating which usually has a high loading of inorganic compounds (particularly metallic salts which are being bioaccumulated). Often fairly brittle in structure, forms sometimes as a senescent biofilm but may still contain an active microbial component. Organic carbon content may decline to less than 0.5% of the total mass.

Extracellular polymeric substances (ECPS), those substances which are synthesized within a microbial cell and promptly excreted to form a tight capsule, loose sheath or generally dispersed slime/gelatinous matrix within which the cell continues to metabolize and reproduce. The collective growth within a biofilm formed by ECPS is sometimes referred to as a glycocalyx. Under some circumstances (e.g. biofilm sloughing or disintegration) the ECPS can become dispersed into the aquatic phase as biocolloids.

Facilitative flow, microbially induced flow which exceeds the rate observed under nonfouled (control) circumstances.

Facultative, a qualifying adjective indicating that the organism is able to perform in the presence or absence of the environmental factor being defined.

Fatty acid composition, a common term for the description of the microbial identification systems that use the relationships of the methyl ester fatty acids (MEFA) to each other within the composition of an unknown microbial cell culture to determine the most likely genus and species. These ratios can be used to accurately identify bacteria at the species level with a variation of less than 0.1% of any given fatty acid occurring within a species.

Fecal coliforms, see *Escherichia coli*.

Fixed film reactor, a system in which a chemical reaction is perpetrated by a biofilm actively growing in a fixed surface position within the reaction vessel that has been selected to perform that specific chemical function.

Food infection, a gastroenteric illness caused by the ingestion of microorganisms which then grow and initiate clinical symptoms of disease in the consumer.

Food intoxication, a poisoning of the consumer by microbial toxins produced in the food prior to consumption.

Food poisoning, a general term which references a gastroenteric disease involving the consumption of food contaminated with pathogens and/or their toxic products.

Gallionella, a gRAM-negative vibrioid to straight rod cell which excretes a ribbon-like twisted stalk out of one side of the cell. This stalk is rich in ECPS polymers and the oxides and hydroxides of ferric, manganic and other metallic elements. Is widely believed to be a dominant component in iron-related bacteria consortia.

Generation time, the time required for a microbial population to double the number of cells through reproduction. Ranges in time from as short as twenty minutes to as long as a number of days, weeks or even years.

Genome, the complete set of genes (hereditary units) present in an organism.

Genus, a group of related species.

Gradients, the defined change in concentrations of a substance which can occur over distance.

Gram, Dr. Christian, inventor (1884) of the most successful differential stain for separating bacteria into two major groups (gRAM-negative gRAM-positive). The gRAM stain is universally applied to determine the Reaction, Arrangement and Morphology of bacteria. In that respect it can be referred to as the gRAM staining technique.

Groundwater Microbiology, the study of any aspect of the phenomenon of microbial activities in and/or at the interfaces to groundwater systems at any depth within the crust. Major emphases at the present time include the biodegradation of pollutant plumes (bioremediation), well plugging and the assessment of the hygiene risks associated with waterborne infestations associable with groundwater sources.

Halophile, organism requiring high levels of salt (NaCl) for survival and growth.

Healthy carrier, an individual who harbors an infectious organism but does not display any diagnostic symptoms.

Heterotrophic bacteria, those bacteria which are able to utilize organic materials as the principal sources of energy and carbon for survival, growth and synthesis.

Hydrolysis, the breakdown of a polymer into smaller structures (usually monomers), by the addition of water, digestion.

In vivo, occurring within the body of a living organism or under natural conditions.

In vitro, occurring in culture under laboratory conditions.

In situ treatment, chemical and/or biological treatments that can be carried out at site in a particular environment (i.e., at the same position or site as the original process to be treated is functioning.

Incubation, the period of time wherein an organism is

grown under suitable conditions. For microorganisms, the dominant factors are the culture medium, the temperature and the composition of the gases surrounding the culture.

Induction, a process in which an enzyme is synthesized in response to the presence of an external substance (i.e. the inducer). Once synthesized, the enzyme will react with the inducer, usually causing degradation. This process does take time which leads to a delay before the reaction can occur (induction period).

Infiltration intake, that point at which there is an infiltration (recharge) into a subsurface water reservoir through porous media.

Invasiveness, the ability of a microorganism to enter an environment (or host), grow, reproduce and spread throughout the domain.

Iron-oxidizing bacteria, bacteria capable of oxidizing iron to the ferric form, commonly an aerobic event.

Iron-precipitating bacteria, bacteria which are able, during their normal life cycle to cause the precipitation of iron salts inside or around the cell or within the aquatic environment.

Iron-reducing bacteria, bacteria capable of reducing iron to the ferrous form, commonly an anaerobic event.

Iron-related bacteria (IRB), any bacteriologically rich organic slime which, when being generated, is rich in salts of oxidized iron due to bioaccumulation. These growths tend to be predominantly orange, red to brown. Color can turn to black under anaerobic conditions due to the activities of SRBs. Also may be defined as any bacterium which can utilize iron within its life cycle in a defined manner whereby the oxides and hydroxides become either bound within the cell, within the ECPS either as a defined appendage, within the general capsule/sheath, or are precipitated in the environment around the growth. An alternative title to the traditional use of iron bacteria.

Koch's postulates, a set of rules for proving that a microorganism causes a particular event (such as a disease or degradation occurrence).

Lag phase, that period which extends from the inoculation of a population into suitable cultural conditions and the start of growth. Can become extended by an induction period.

Legionella, a small gRAM-negative aerobic rod which is able to grow over a broad spectrum of temperatures and pH values. Sometimes occurs in water distribution systems where the particulate organic carbon is above 0.6 mg/L. It also occurs in the intake/output lines from water heaters and is associated in waterborne infections causing Pontiac fever and Legionnaires' disease.

Lipids, alternative term to fats, water insoluble molecules.

Low nutrient environment (Oligotrophic), an environment within which the nutrients (C,N,P and S) are either collectively very low or where one is sufficiently low as to be limiting the microbial growth within the system. It should be noted that many "low nutrient environments" may in fact support very heavy biofilm growths since the biofilm will act as a bioaccumulator of the nutrients even in aquatic systems where the dissolved nutrient levels are observed to be very low.

Macrofouling, extremely intense forms of biofouling in terms of both activity levels and scale.

Magnetotactic bacteria, bacteria which are able to position themselves within magnetic fields. Specific magnetosome particles within the cell appear to be responsible for this orientation.

Marginal plugging, well/groundwater fouled interface which has become subjected to sufficient biofilm formation to influence (i.e. reduce) the rate of flow of water into the water well. The level of flow reduction that may be expected with a marginal plugging would be less than 20%.

Mechanical disruption, the use of abrasive, rapid pressure fluctuations, ultrasonics or rapid temperature changes from freezing through to steam generation to cause a mechanical disruption of a slime plug.

Membrane filter (technique), the use of a thin filter membrane to filter out (to produce a sterile filtrate) or to concentrate (on the surface to incubate and enumerate colonies that may grow) microorganisms present in an aqueous sample.

Mesotroph, any organism able to grow strictly within the temperature range of 15°C to 45°C.

Methylene blue, a crystalline dye used to determine microbial activities by enzymic activity which causes the color of the dye to change from blue to colorless. The rate at which this reduction occurs can be linked to the amount of microbial activity. Commonly used in the dairy industry to indicate the microbial populations in milk.

Microbial growth potential, the potential for any given environment to support the growth of microorganisms within the system based on the known environmental parameters for that system. For example, a high nutrient environment with a mesotrophic temperature and pH of 7.5 with saturated oxygen would be considered to have a high microbial growth potential.

Microcosm, a small portion of an environmental system that is separate and operates as a distinct biological entity. Usually involves very small volumes where conditions are optimized for the growth of a particular species or consortium of microorganisms.

Model water wells, laboratory-scale replications (microcosms) of water wells wherein the flow rates, nutrient loadings, constructional methods, treatment systems and microbial biofouling can be rapidly replicated and studied.

Most probable number (MPN), a statistical method for the prediction of a microbial population in a liquid by the presence/absence of growth in various dilutions.

Motility, the property of directional movement of a cell under its own power. Not to be confused with Brownian movement in which there is a random oscillation of the cell as a result of external influences.

Negative staining, a staining procedure in which the dye forms a dark background while the specimen remains unstained. Convenient for examining the outline and arrangement of cells.

Nitrification, the aerobic conversion of ammonium via nitrite to nitrate. Performed solely by the nitrifying bacteria which are usually classified using the prefix *Nitro-* or *Nitroso-* depending upon the function performed.

Nosocomial infection, an infection which is acquired and develops within a hospital or clinical care facility environment.

Nosocomial bacteria, bacteria which occur naturally in the environment but can under some circumstances, cause infections in a suitably weakened host.

Nutrient depletion strategy, the deliberate restriction of one or more nutrients from an environment in such a way as to attempt to control the growth of microorganisms within that system.

Obligate a descriptive adjective referencing environmental factors essential to the survival, growth and reproduction of an organism.

Oligotroph, an organism able to flourish in an environment having a low-nutrient regime.

Pandemic, a worldwide epidemic.

Partial inhibition, the restriction of microbial activity by interference with the metabolic and/or reproductive functioning of the organism by manipulating the environment (e.g. the addition of a low concentration of disinfectant). It is intrinsically implied that the bulk of the microorganisms will eventually recover after a period of repair and adaptation.

Particulates, suspended inorganic or organic defined materials formed into coherent masses with a diameter in excess of 0.5 microns. These particulates vary in diameters of up to 250 microns or more depending on flow characteristics. The origin may be from the sloughing of biofilms, direct chemical reactions within the water, planktonic microbial growths and any combinations of these.

Pasteurization, the process of heating a heat sensitive substrate to just a sufficient extent to reduce the nuisance microbial population to an acceptable and manageable level. Usually performed within strict temperature and time constraints (e.g. 60°C for 30 minutes).

Petri dish, a shallow glass or plastic vessel with a fitted lid. It allows microorganisms to be grown under aseptic (free from viable contaminants) conditions usually in agar culture media.

Planktonic, a microbial community growing as independent cells while suspended within the water and moving freely on the currents.

Plasmid, an extrachromosomal genetic element which is not essential to the growth of the organism but may contain information relating to unique additional functions.

Plugging, the phenomenon of an intense microbial biofilm growth causing such a heavy slime generation that there is a significant (greater than 20%) loss in flow (hydraulic conductivity) through the interstitial spaces of a saturated porous medium. Used synonymously with the term clogging.

Plugging risk index (PRI), the specific utilization of the microbial growth potential within an environmental system to predict the rate at which biofouling and plugging are likely to occur. Factors to be taken into consideration include the basic physical and chemical parameters along with an evaluation of the likelihood of microbial infestations occurring.

Preventative management mode (PMM), the manipulation of such factors and changes as may be deemed necessary to prevent the onset of a physical/chemical microbial problem.

Pseudomonas species, strictly aerobic gRAM negative rods which occur widely in groundwater systems particularly where oxygen and/or nitrates are present in sizeable concentrations, associated frequently in IRB communities, major contributor to biodegradation events. Several species are associated with nosocomial infections.

Pseudomonas aeruginosa, a gRAM negative strictly aerobic rod-shaped bacteria which occurs widely in the environment and can cause severe wound and post operative infections.

Psychrotrophs, those organisms that are able to grow at temperatures below 15°C. They are differentiated into two groups. The first group are the obligate (or strict) psychrotrophs and will grow only over a range below 15°C. The other group are the facultative psychrotrophs which can grow usually much faster at temperatures above 15°C.

Recolonization, reestablishment of a microbial population (sessile and/or planktonic) within a given environment from which the population had been either eradicated or lost.

Rehabilitation, the process of restoring an environment (such as a water well or groundwater system) to its former (unfouled, pristine) state by the utilization of various control strategies. Restoration is an alternative term widely used.

Reinfection, the occurrence of a fresh infestation in a given installation by organisms either implanted by the maintenance crews or arriving at the water well after passage through the groundwater system or from surface recharges.

Repositioning, the strategy for avoiding a recognized infestation risk by the relocating of plant equipment and water well installations away from the infestation zone.

Robbin's spool, a patented device which can be implanted into the walls of a given transmission or storage system for liquids in order to monitor the rate of biofouling occurring by the patterns of colonization that occur. The device provides a mechanism for removing the colonizing biofilm from the system for evaluation and testing to diagnose the causant agents.

Roll over, the positioning of a small part of a transmission pipe for liquids (usually oils) at a lower level in order to build up turbulence and water activity to exaggerate any possible biological activities occurring within the pipeline as a whole (worst-case scenario).

Scanning electron microscopy (SEM), the process of microscopy that uses reflected electron beams to form an image similar to that on the surface of any given material. By this means, it is possible to view biofilm formation, corrosion and the effect that this is having on surfaces without destroying the surface growths.

Sequential mechanisms, the utilization of a number of mechanisms in a specific sequence in order to ensure the maximum possible desired effect.

Sessile, growing on a surface attached to or incorporated into a biofilm or some other structure which prevents the organism from becoming rapidly detached into the liquid phase.

Sheath, a hollow tube-like structure surrounding a chain of cells. Bacteria bearing this feature are sometimes referred to as the "sheathed bacteria".

Shock chlorination, the act of applying large doses of a form of a chlorine (e.g. >5,000 ppm of sodium hypochlorite) in order to maximize the disinfectant activity and reduce the presence and activity levels of the incumbent organisms.

Slime, a jelly-like substance rich in ECPS often incorporating granular material. Usually forms a coherent film covering walls, filling the interstitial spaces within the porous media such as that of aquifers, or coating well screen slots in way to cause a degeneration in the quality of the postdiluvial water through increased turbidity and decreases in the quantity of water flowing through a given system.

Slime bacteria, those bacteria which reside and reproduce within a slime. The unique feature here is that the slime is capable of directed movement and the resident bacteria usually feed on other (commonly gRAM-negative) bacteria or cellulose.

Sloughing, the act of a biofilm (through casual events or as a result of flow, maturation or pressure changes) disintegrating. This sloughing may cause layers of the biofilm to move to the suspended particulate phase detached from the rest of the biofilm. Such sloughing can be locally reassimilated by downstream biofilms or remain suspended leading to a reduction in the quality of the product water.

Spreadplate, a technique where the bacteria to be enumerated are spread out over the surface of a suitable agar culture medium with a minimum of trauma. This is considered to heighten the probability of recovering and growing the incumbent organisms as distinct colonies.

Stabilization, the development of an environment within which the dynamics of the biological activity becomes stabilized causing no greater nor lesser amount of problems over that period of time.

Standard plate count (SPC), the recommended procedure for determining the number of bacteria (as colony-forming units, cfu) in water utilizing high mesotrophic temperatures. This method is generally used as a backup system for insuring a low risk of nosocomial bacteria within a given water system.

Streakplate, a specific method for streaking a diluent over the surface of an agar plate to allow convenient enumeration of subsequent colonies that may be formed upon incubation. (See also spreadplate).

Sulfate-reducing bacteria (SRB), anaerobic bacteria able to reduce sulfate with the release of potentially corrosive hydrogen sulfide.

Thermal death point, the lowest temperature required to destroy a microbial suspension in ten minutes.

Thermal death time (TDT), the shortest period of time required to kill all of the organisms in a microbial population under specified conditions (i.e., temperature, cell age and number, volume, heat input and carrier substrate).

Thermotroph, those bacteria which are able to grow only at temperatures in excess of 45°C.

Total aerobic plate count, an evaluation of the number of aerobic (oxygen-requiring or aerotolerant) bacteria within a given system by extinction dilution techniques using spread plate techniques and temperature/incubation times optimized for the maximum population assessment.

Total coliforms, gRAM-negative facultatively anaerobic, oxidase-negative, fermentative enteric bacteria which are either directly or indirectly associated with the pollution of waters by fecal material.

Total nitrogen (TN), an analysis for nitrogen within water. Fractions include nitrate-nitrogen, ammonium-nitrogen and Kjeldahl nitrogen. While the former states are considered soluble, the latter state relates to nitrogen bound into various forms in biological material.

Total organic carbon (TOC), the total organic carbon recoverable in a sample. Fractions may include soluble utilizable organic carbon (SUC), soluble nonutilizable organic carbon (SNUC), particulate utilizable organic carbon (PUC) and the particulate nonutilizable organic carbon (PNUC). The ratios of SUC: SNUC and PUC:PNUC will give an indication of the degradability of organic pollutants within the system.

Total phosphorus (TP), the total amount of phosphorus recovered from an aquatic system including the soluble inorganic phosphorus (SIP), particulate inorganic phosphorus (PIP), particulate organic phosphorus (POP) and the soluble organic phosphorus (SOP). The interactive ratios between the SIP:PIP:POP and SOP can be used to determine the amount of biological activity occurring.

Treatment fringe effect, the phenomena generated on the fringes of a given physical/chemical/biological treatment zone. At the lower treatment dosages occurring at the fringe, effects may be seen wherein the microbial growth may become partially suppressed, stimulated or be unaffected by these particular factors.

Tubercles, raised encrustations on a surface within which a biofilm may be generated. Frequently associated with electrolytic corrosion.

Tyndallization, the suppression of a microbial growth and the repeated application (normally three) of pasteurization in which sufficient time is allowed between treatments for the surviving organisms to grow to a point of becoming more vulnerable to heat (i.e., the next treatment).

Ultramicrobacteria (UMB), microbial cells that have undergone severe environmental stress and have reacted by entering a phase of suspended animation. Where this happens, much of the cell contents and water are either expressed or utilized so that the diameter of the cell shrinks to between 0.1 and 0.5 microns. As a final action to this severe stress, the cells will reproduce. These will remain fully viable and will grow again rapidly when the environment conditions are appropriate. They are unable to attach to surfaces.

Viable counts, the enumeration of microorganisms specifically restricted to those cells known and recognized to be viable.

Vyredox, the trademark for a patented system for reducing the plugging of water wells and/or improving the water quality by oxidizing out the iron and manganese *in situ* away from the well screen. The method effectively shifts the redox fringe back into the aquifer by creating a larger oxidative zone.

Waterborne infections, generalized name applied to those infections created by pathogens which have reached the host via water.

Well pasteurization, utilization of high temperatures (in excess of 60°C for 30 minutes) in order to destroy the biofouling formed by microorganisms at the water well/groundwater interface.

Wolfe's medium, a specialized selective medium for the growth of *Gallionella* used in the diagnosis of iron bacterial infestations within wells.

WR medium, a modified form of Winogradsky's medium developed at the University of Regina (hence WR) which appears to be able to enumerate the bulk of the iron related bacteria.

11

Abbreviation Definitions

%pp, pigmented particulates estimated by reflectivity
AI, aggressivity index
ap, achromogenic (non-pigmented) particulates
APIRB, averaged population of iron related bacteria
BAQC, biofouling assessment quality control
BCHT, blended chemical heat treatment
BHI/4, brain heart infusion agar at one quarter strength
bi, biological interface
BIB, biological intercedent barrier
bp, background particulates
BQ, before quiesent phase
BTX, benzene, toluene and xylene
bv, biofouled volume (around a water well)
B_z, biofouling severity index
cbw, theoretical width of a cylindrical biozone
CC_d, colony count for dilution (d)
CC_v, MF correction volume (v) factor to 100 ml base
CD, Czapek-Dox culture medium
cfu, colony-forming units
CHI, consortial heterotrophic incumbents
CLF, coliforms determined using M Endo LES agar
cw, causal water

cwc, concentration in causal water
dd, days of delay
D_{min}, minimal distance from a fouled well for installation of a new well
D_{zbz}, theoretical diameter of biozone around a well
DB, days to bleaching
ECPS, extracellular polymeric substances
EDB, extinction dilution diodetection
ELES, LES Endo agar
ETO, ethylene dioxide
f., factorial
FC, fecal coliforms
$FC_?$, factorial comparison of two groups of bacteria (?)
FID, floating intercedent device used in the BART™ and MOR systems
FR, flow rate
FS, fecal streptococci (enterococci)
ft, filtration time
G-COLI, broad spectrum of coliforms (general)
GAB, gross aerobic bacteria determined using BHI agar
gRAM, Gram reaction, arrangement and morphology
HB, heterotrophic bacteria determined using R2A agar
hd, hours of delay (applicable to the COLI-MOR test)
IA, immersed attachment
id, incumbency density
ID, inside diameter (of a biozone)
IGU, incumbency for gravimetric units
IPP, incumbent possible population
IRB, iron related bacteria
IVU, incumbent viable units
IWID, in-well incubation device
MAD, microbial aggressivity determination
MF, membrane filtration
MF_c, correction factor applicable to MF determination of cfu
MIC, microbially induced corrosion

MIF, microbially induced fouling
MGG, microbial gas generation
MLC, molds determined using Czapek-Dox agar
MLP, molds determined using PDA agar
MOB, mettalo-oxidizing bacteria
MOR, measurement of risk
MPA, microbial particle assessment
MPN, most probable number
MPNB, metallo-precipitating non-oxidizing bacteria
N/D, not detected
OD, outside diameter (of a biozone)
OV, occupancy volume (for a biofilm)
p/a, presence/absence
P-A, presence-absence, see also p/a
pa, porosity factorial for aquifer
PCB, polychlorinated biphenyls
PD, potato dextrose agar
PERP, period of enhanced releases of particulates
pg, porosity factor for pack material (e.g., gravel pack)
pp, pigmented particulates
PP, possible population
pr, pumping rate (from a water well)
PSE, pseudomonads determined using AA agar
PSE, Pfizer selective enterococcus agar
pw, postdiluvial water
pwc, concentration in postdiluvial water
q, quantitative
RE, rate of elevation of water in an IWID being charged
rif, radius from well mid-point to inner edge of biozone
RX, reaction pattern in sequence (x)
rxf, radius from well mid-point to outer edge of biozone
s/q, semi-quantitative
S-COLI, narrow spectrum of fecal coliforms (specific)
sp, sheared (sloughed) particulates
SPBR, spreadplate bacterial population relationships

SRB, sulfate reducing bacteria
STR, streptomycetes determined using R2A/10 agar
T.F.T.C., too few to count
T.N.T.C., too numerous to count
TAB, total aerobic bacteria
TC, total coliforms
THM, trihalomethanes
tp_c, time of pumping to the conclusion of evidences from a biozone
tp_i, time of pumping to the initial presence from a biozone
TSA, triple sugar iron agar
u/d, uptake and/or degradation
VB, volume of the biofilm
V_{bz}, volume for a given biozone
vp, volume suspended particulates per mL
vu, viable units
VV, void volume
WR, iron related bacteria determined using WR agar
WRF, white rot fungi
Y, magnifying factor applicable to a CC_d to calculate cfu
Z, number of acceptable CC_d values used in calculating cfu

Appendix One

Using the File Storage and Interpretation Program BART™ SOFT Version 1.1

A Program to Record and Interpret BART™ Tests

This program is available on the floppy disk provided with the supplementary software package. It is set to operate in MS-DOS with a minimum of user installation. Files are created for each test performed in ASCII format which is Lotus 1,2,3 compatible. It is **essential** when running this program to have the **computer clock set and operating** with reasonable accuracy (plus or minus one minute) since the calculation of the days of delay is based upon the computer clock time. Read the instructions for operating the program prior to using the program.

1. System Requirements

To run the BART™-SOFT program, the minimum need is for any IBM-PC compatible computer running MS-DOS version 3.00 or higher. There must be at least one floppy disk drive; for improved speed, a hard disk is recommended. A basic familiarity with MS-DOS is essential, reference can be found to the necessary knowledge base in section A below. In particular, it is essential to know how to:
- list a directory.
- copy, delete, and rename files.
- create and delete subdirectories.
- change the current directory and drive.
- name a particular file or set of files in any of the above commands.

1.1 Manual Conventions

In this manual, several conventions have been adopted for enhancing the clarity of the explanations. Text should be typed in exactly as shown:

Function keys, and other special keys are printed in uppercase, e.g., F1, F10, PGUP, INS, ENTER.
Key combinations are printed as a list of key names with plus signs in between them. For example:

CTRL + c   means press the control key, hold it, then press c.

ALT + f   means press the alt key and hold it, then press f.

2. Instructions for Running the Software
2.1. Starting Up

Boot the computer and make sure that the clock is set to the correct time. Failure to set the clock properly will seriously reduce the convenience of the program since all chronologically based data will have to be manually entered. If your computer system does not have a built-in real-time clock, you will have to set the software clock every time you boot the computer. The commands to do this are date and time.

> A:\date ENTER
> A:\time ENTER

Type in the date or time in the precise format shown on the screen. Failure to do so will cause the entry to be rejected.

Run the BART™ program by typing bart.

> A:\bart ENTER

A title screen with the program's name and version number will appear. Hit any key to clear this title screen and commence the program. Here, a main menu bar will be displayed with the headings **File Edit Query Report**.

Each of these menus is activated by pressing the ALT key and the first letter of the desired menu, e.g., ALT-F will pull down the File menu. When this happens, a set of possible functions hangs down on the screen with activated heading. Alternatively, pressing **F10** will cause the most recently activated menu to again be pulled down.

2.2. The Pull-Down Menu System
2.2.1. The File Menu

The File menu consists of the following entries:

Open
Save
saveAs
Close
Pack
Quit

Open will normally be the first command to invoke when starting the program. Either type in the letter highlighted and press ENTER or place the cursor using the cursor keys to the right of the keyboard over Open and press ENTER. The Open command will open an existing database in the current directory (last one worked on); it is possible to create a new database. The program will request the name of the database to open or create, at which time, type in the desired database name or press F1 to display a menu of existing databases held in storage on the disk. If there are two disc drives with the program disk in A: drive, a formatted file disk can be used in B: drive to store data on. Where this is the case, type in B: before the name of the directory you wish to recover.

The name chosen becomes the "current database"; the database on which the data will be filed and interpreted. The program displays the first record entered into the database, or a blank file form if the database is empty.

The Save command saves the current database to disk with any changes that have been made to the file(s) form. This Save command is automatically invoked when the Close or Quit commands are used. This reduces the chances of accidentally losing data.

The saveAs command is very similar to the Save command but there is one very major difference. It allows you to specify a new name to save to. This allows a copy of the current database to be saved into a newly created database which now automatically becomes the current one being used by the computer.

The Close command saves the current database to disk and clears it from memory, ending a particular routine. A new database may now be opened or created.

The Pack command permanently removes any records that have been pre-designated for deletion. When records are designated for deletion from a database, they are not immediately removed, but are merely marked as "deleted". The Pack command can be used when appropriate to physically remove those records from the files and so recover some disk space for future use.

The Quit command first saves the current database in the computer, closes out the program and returns to DOS. There is a prompt to confirm this request before the computer will enact the abortion of the program. The screen will show:

>A:\

when the program has been erased from the memory.

2.2.2. The Edit Menu
The Edit menu consists of the following entries:

    Add
    Head
    Tail
    Next
    Prev
    Edit
    Delete
    Undelete

The Add command adds a new, blank file form (record) at the end of the database to record the information on a new BART™. It also allows information to be entered. After entering the information, you save the record by moving to the last part of the illuminated cursor field, and pressing RETURN. You can move to this last field either quickly by pressing CTRL-END; repeatedly pressing the TAB key; or by repeatedly pressing the DOWN cursor key.

The Head command moves you to the first file form record in the database, while the Tail command moves you to the last record in the database. These two functions can also be performed by the HOME and END keys.

The Next and Prev commands move forward and backward to the next file form and the previous file form records, respectively. These two functions can also be performed by the PGUP and PGDN keys.

The Edit command allows you to edit of any of the information stored in the fields of the current record. The F9 key performs the same function as the Edit command. The basic operations that can be performed while editing are:

1. Entering new information and modifying old information. Simply typing text in a field will enter the text into that field. By default the text will overwrite any previous information in that field. Entering the insert mode by pressing the INS key, will cause existing text to be displaced to the right as new text is entered. The cursor becomes larger to indicate that you are in insert mode. Pressing the INS key again returns the program to the normal overwrite mode. Some fields have restrictions on the type of information that can be entered into them. For example, alphabetic text cannot be entered into a field which expects a number, and vice versa. If an attempt is made to enter the "wrong" type of text into a field, the computer will beep to indicate the error.

HOME - moves you to the first character in the current field.

END - moves you to the last character in the current field.

LEFT - the left cursor key moves the cursor one to the left.

RIGHT- the right cursor key moves the cursor one to the right.

DEL - the delete key deletes the character at the cursor position.

BS - the backspace key deletes the character left of the cursor.

INS - toggles between the insert/overwrite mode.

2. Moving from one form file field to another:

RETURN - completes the editing of the current field and moves to the next one. When the last field editing process is terminated and the record has been saved, the next field can be entered.

DOWN - the cursor down key moves to the next field in the same form file. If on the last field of the form file, the cursor moves to the first field of the same file. This function can also be performed by the TAB key.

UP - the cursor up key moves to the previous field on the file form. If the cursor is on the first field, it will automatically move to the last field. This function is also performed by the SHIFT-TAB key.

The Delete command will delete the current record. Press the F4 function key when the record is displayed as the current file form, then the instruction to delete the record will appear.

The Undelete command will undelete the current record. Pressing the F4 function key when a deleted record is displayed will protect the record from deletion.

2.2.3. The Report Menu

The Report menu consists of the following entries:

Print all

Record print

The Print all command allows the entire database to be sent to the printer, or to a disk file. To send the database to the printer, simply use the special filename "PRN" as the destination. For example, you would print the database to a file named "results" with the following sequence (type in, consequence):

ALT + r	Pulls down the Report menu.
p	Chooses the "Print all" menu item.
results	Specifies the file name to print to.
ENTER	Confirms the command.

The Record print command allows the current record to be sent to the printer, or to a disk file. To send the record to the printer, simply type in the special filename "PRN" as the destination. For example, you would print the current record to the printer with the following sequence (type in, consequence):

ALT + r	Pulls down the Report menu.
r	Chooses the "Record print" menu item.
prn	Sends the file to the printer.
ENTER	Confirms the command.

2.3. Description of Fields

Sample Name: The name of the BART™ sample being tested. This will generally be the name of a site where a series of tests is to be performed. For example, if all the BART™ tests were performed at a site called WELL23, the sample name could be WELL23, and would be differentiated by their individual sample numbers, given in the next field. The sample name may be up to 14 characters long.

Sample Number: The series number assigned to this

particular sample within a series. The number may be up to 5 digits long.

BART™ Type: This will be the code name applied to a particular test, such as an IRB- or SLYM-, up to 5 characters long. The interpretation of the data will vary depending upon the entry made into this field. Care should therefore be taken to ensure the correct letters (three or four in length) are used.

Taken At: This 14-character field is used for storing additional information describing pertinent conditions under which the sample was taken.

Example 1:

If BART™ samples were taken at 5 minute intervals for a one hour period, this field could indicate the time as: 5 min, 10 min, 15 min, etc.

Example 2:

If BART™ samples were taken along a stream bed at 100 foot intervals, this field could indicate the distances as: 100 feet, 200 feet, 300 feet, etc.

Start Date: This field is automatically filled by the computer when the record is created. The current date and time is read from the computer system clock. The start date may also be manually changed by the user.

Last Update: This field is automatically filled by the computer when the record is modified in any way. The current date and time is read from the computer system clock. The last update may also be manually changed by the user.

*dd*1: This field stores the days of delay (*dd*) to the first reaction. This will be a number from 0.1 to 99.9. The special value 0.0 indicates that no reaction 1 has yet occurred.

RX1: This field stores the numeric type of the first reaction. This can be a number from 1 through 99. The special value 0 indicates that no observable reaction has yet occurred.

*dd*2: This field stores the days of delay to a second and distinctive reaction. This will be a number from 0.1 to 99.9. The special value 0.0 indicates that no secondary reaction has occurred yet.

RX2: This field stores the numeric type of the second reaction observed at *dd*2. This will be a number from 1 through 99. The special value 0 indicates that no secondary reaction 2 has occurred.

*dd*3: This field stores the days of delay to the third reaction. This will be a number from 0.1 to 99.9. The special value 0.0 indicates that no reaction 3 has occurred.

RX3: This field stores the numeric type of the third reaction. This will be a number from 1 through 99. The special value 0 indicates that no reaction 3 has occurred yet.

Comment: This field can be used to store an arbitrary comment relevant to that particular test or sample, up to 50 characters in length.

2.4. Calculated Interpretation Fields

These fields are not actually stored in the database file but are calculated from other existing fields and presented on the screen for the user's information. The calculated interpretation fields currently used are:

1. **Current *dd*** - The number of days since the record was created. Note that when a reaction first occurs, you must enter the current days of delay in the appropriate field (*dd*1, *dd*2, or *dd*3) and the associated reaction number in its appropriate field (RX1, RX2, or RX3). The Current DD field contains the value that should be entered in the DD1, DD2, or DD3 field.

2. **Possible Population (*pp*)** - is an estimate of

microbial populations in the water sample based on the speed with which the reaction occurred within the particular BART™ Type, and the DD1 and RX1 fields. The population is expressed in colony-forming units per milliliter (cfu/mL). This is calculated using the moment method, gamma distribution involving two constants (alpha, beta) for the IRB-, SLYM-, SRB-, and FLOR- BART™ tests. Confirmation of these data so obtained can be obtained by duplicating the testing using standard procedures.

3. A verbal description of the activity of the population is characterized as highly aggressive, moderately aggressive, significant population, some detected, or none detected.

4. **Sample Session**
   This section illustrates a sample session with the BART software in which a database is created using one IRB-BART™. After three days have passed, the the IRB BART goes positive. The reactions (RX) was observed to be type 5. Reaction type 5 is an accumulation of small yellow-brown gas bubbles under the FID, indicating the presence of anaerobic gas-generating iron related bacteria. To prepare this record, several fields have no information in them. TAKEN-AT was not needed for it is designed to be used with a multiplicity of tests being conducted in the same test series. Here, the SAMPL-NUM field is sufficient. $dd2$, $dd3$, RX2, and RX3 are zero (which means the same as blank) because there has not yet been any reaction 2 or 3. The comment field was left blank because there was no exceptional information to be retained other than a standard IRB reaction 5. Based on the BART Type, the $dd1$, and

RX1, the Possible Population is estimated at 95,000 cfu/mL. The last line of the record gives a qualitative judgement of "significant" bacterial activity level.

The type of screen display which would be presented when starting a file form field for the above reaction would allow the basic information to be entered and the timing of the test would be initiated by the acceptance and storage of the file form. At the time of starting the test, the display would be:

File   Edit   Query   Report
                         Current Database: TRIAL.DBF
                         Current Index:
Record: 1/1
SAMPL_NAME : TEST   SAMPL_NUM   : 1
BART_TYPE   : IRB   TAKEN_AT   :
START_DATE : 1990-10-09 16:50
LAST_UPDAT : 1990-10-09 16:50   Current *dd* : 0.0
*dd*1   : .   RX1   :
*dd*2   : .   RX2   :
*dd*3   : .   RX3   :
COMMENT   :
    PP:

F1-Help F3-Add F9-Edit F10-Menu

Each day the test is observed and, if the file form is brought up, it will be noted that a current *dd* (day of delay) will be displayed set by the computer clock. When a reaction is noted, this *dd* is typed in as *dd*1 and the reaction type (RX) typed in on the same line. The file form will now display a possible population (PP) based on the data and interpret the aggressivity of the incumbent microorganisms.

File   Edit   Query   Report
Current Database: TRIAL.DBF
Current Index:

Record: 1/1

SAMPL_NAME : TEST   SAMPL_NUM : 1
BART_TYPE   : IRB   TAKEN_AT   :
START_DATE : 1990-10-09 16:50
LAST_UPDAT : 1990-10-12 17:11   Current dd : 3.0
dd1   : 3.0   RX1   : 5
dd2   : 0.0   RX2   : 0
dd3   : 0.0   RX3   : 0
COMMENT   :
PP: 95 thousand cfu/mL
Significant iron related bacteria population.

F1-Help F3-Add F9-Edit F10-Menu

A. Using MS-DOS

1. File System.

Information is stored on disk drives in the form of files. Disk files are of basically two types: ordinary files and directories. Examples of ordinary files are data files created with word-processing, spreadsheet, database and drawing programs; or files containing the operating software programs themselves (i.e., word-processors, spreadsheets, database programs, etc.). For instance, these two files can be designated:

123.exe - represents the Lotus 1,2,3 software program.
sales.wk1 - a worksheet data file with sales information.
Both are ordinary files.

Directories, on the other hand, is a repository for retention of files and so performs a very different function entirely. A directory can hold both ordinary and directory files. These files are considered to reside inside the directory. The convenience of directories is that they provide a convenient means of organizing a large numbers of files into a series of logical categories. This is in much the same way that a filing cabinet allows the grouping of a large number of paper documents into folders. Because a directory can even hold other directories as well as ordinary files, the file system can be structured almost like a tree. A single main directory forms the root with subdirectories forming the branches.

In MS-DOS, every disk has at least one directory: the "root" directory, named \. In the case where only floppy disks are used, that is probably the only directory needed since each disk will hold too few files to allow any complex organization. On a hard disk, however, it is possible to hold thousands of files. It is here that a directory structure becomes essential to manage those files.

The information presented in Appendix One can also be found in several commercially available reference books on MS-DOS. Virtually any of these books, which can be obtained from most book stores, will give you the information you need to start using MS-DOS.

2. Listing a Directory.

The dir command is used to list the files in a directory. The command can take one request which specifies the directory to list. At the prompt (>A: or >B:) type in the request for directory information using "dir" and ENTER. If there are only specific files to be listed this can also be requested:

dir

Lists all files in the current disk drive directory.

dir b:

Lists all files in the current directory of drive b:.

dir c:\business\letters

List files in the given directories.

dir c:\business*.wk1

List worksheet (.wk1) files in the business directory only.

Note that c: would usually indicate a hard disk storage, b: a second floppy disk file.

3. Copying, Deleting, and Renaming Files.

Copying files is done with the "copy" command, deleting files with the "del" or "erase" command, and renaming files with the "rename" or "ren" command. You cannot rename directories. At the prompt (e.g., >A:) type the command required.

copy sales.wk1 old.wk1

Make a duplicate of sales.wk1 in the current directory with the name old.wk1.

Note that sales.wk1 remains on file.

copy *.* a:

Copy all files in the current directory to drive a:.

del \business\letters\a*.*

Delete all business correspondence files whose names begin with the letter "a".

ren payroll.wk1 pay.wk1

Renames a file in the current directory named payroll.wk1   to pay.wk1.

4. Creating and Deleting Subdirectories.

The command mkdir or md will make a new directory and the command rmdir or rd will remove a directory. You cannot remove a directory unless it is empty.

md newdir   Makes a directory under the current one.

md c:\newdir   Makes a directory under root of drive c:

rd junk   Removes a directory named junk under the current one.

rd c:\newdir Removes the newdir directory under the root of drive c:

5. Changing the Current Directory and Drive.

Most commands that manipulate directories and ordinary files assume a default or current directory and drive. That is why when typing dir with no further instruction, a listing will be shown for the current directory and drive. Changing the current directory or drive is done as follows:

cd \123   Changes to the directory named \123.

cd ..   Changes to the parent directory; i.e., the directory one level up from the current one.

a: Makes drive a: the current one.

6. Naming Files.

A basic filename consists of a basename with an optional extension. The basename may normally have up to 8 characters, while the extension may have up to 3 characters. The basename and extension are always separated by a period (.). Filenames may use just about any characters except these: * < > ¦ \ / " ' . ?
It is perhaps best to stick to using just letters and digits. Here are some examples of legal and illegal filenames:

Legal names: 123   abbie.doc
Illegal names:

Problem.   Do not use a period
bar¦none   Can't use the ¦ symbol.
bad name   Can't use spaces.
help?   Can't use the question mark.
miscellaneous.doc   More than 8 characters in the basename.

The special characters * and ? are called wildcard characters. They can be used in a command to represent a set of matching files. The asterisk represents an arbitrary list of characters, and the question mark represents an arbitrary single character.

a?.* can be used to match with all files that have a two letter basename beginning with an "a", and with any extension or no extension at all.

. matches to all files.

*.doc matches all files with a .doc extension.

At any point where a filename can be used in a command, for example, a dir filename, an optional drive specifier and/or pathname can be present. These override the default or current directory and drive and cause the command to operate on the specified drive and directory.

dir f

Lists the file named f in the current directory.

dir a:f

Lists file named f in current directory on drive a:.

dir c:\business\letters

Lists all files in the given directory.

B. Glossary of terms used with MS-DOS

Boot, An obscure term, meaning to restart the computer by either typing the key combination: CTRL + ALT + DEL; or by turning the computer's power off and on again. This term originates from the expression "pulling yourself up by your own bootstraps", an impossible task for humans but a routine task for the computers!

Database, A single disk file used to store information on, for example, a set of BART™s. Information on the individual tests are stored in records within the database. Field, A single item of information in a record. A record is made up of many fields. An example of a field is the BART™ type, which can have the value IRB, SLYM, SRB, etc.

Record, An item of information describing one BART™ test.

A record contains the following fields:

SAMPL_NAME name of the BART series.
SAMPL_NUM sequential sample number.
TAKEN_AT   additional specifier, indicating time or place.
START_DATE the start date of the test.
LAST_UPDATE   the start date of the test.
$dd1,dd2,dd3$ the days of delay to reaction 1, 2, 3.
RX1,RX2,RX3 reaction types of $dd1$, $dd2$, and $dd3$.
COMMENT   any comments.

Records are stored sequentially in the database, in the same order that they were entered.

Appendix Two

Calculating the Theoretical Radius
of Biofouling (Non-Intrusive Model)

During pump testing to evaluate the possibility of a biofouling event occurring in and around a water well, a number of strategies have been suggested in the text. These range from the "double two", "triple three" and the BAQC-16 procedures, all of which rely on the well being retained in a quiescent state prior to testing for biofouling by chemical, physical and microbiological means. The concept behind these scenarios is that the microbial consortia within the biofouling zones will be stressed by the shifting in environmental conditions (e.g., by no longer pumping). These conditions cause the organisms to detach from the biofilm and to become suspended particulate and planktonic (freely suspended) forms which move with the hydraulic flows. When a favorable environment emerges, these detached particles could reattach and colonize any available econiches.

From the evidences generated to date it would appear most likely that well screens are often surrounded by masses of biological growth which may or may not directly impede the flow of water into the well. The net effect of this would be to reduce production, cause greater drawdowns and adversely affect water quality. In shape, the biomass outside of the well screen may reflect a number of factors. These include the position of any redox fringe (boundary between an oxidative and reductive zone), the flow patterns of the water moving towards the wells screen (biomass extensions will move out to meet in-flowing nutrient pathways), fracture pathways and any tenable porous structures. In reality, these constraints would likely mean that the physical shape of the attached biofouling mass is likely to be irregular and reflect the above constraints.

In the real world, zones of massive "cauliflower" like growths of slime are sometimes seen growing into the wells water column through the screen slots often only at some of the depths. The position of these slime growths indicates zones where the biological activity is maximal close to or even at the screen position. At other sites, the growths may be even more intense but occur further away from the well. This would result in these major biomass formations being not observable from the well's water column. Lack of ability to directly view these covert biofouling structures may give the operators a false sense of confidence.

When well installations have been dug out for re-use (such as around dewatering mine operations), visible slime formations can sometimes be recovered *in situ*. Very often in shallower wells, the major visible biofouling may occur in the shape of an inverted "top hat" with the brim level the uppermost part of the producing zone in the ground water. Other times, the biofouling may be seen as a cylindrical growth of slime close to and often extending some distance from the well screen into the aquifer. Where there is a significant movement of nutrient rich water from one particular direction, finger-like (dendritic) outgrowths can sometimes be found pointing outwards in the direction the water is coming from. If the aquifer is rich in oxygen and nutrients, a very dispersed type of fouling may be noted throughout the formation.

There would appear to be no standard structural form for the biofouling structures that can be found around water wells except these may be composed of a number of concentric biozones. The size of these biozones is, however, measurable by undertaking a post-quiescent monitoring of the diskharge waters from such a fouled well. This quiescent period would form during which the detached or easily sheared particles from the intrinsic stressed biofilms would subsequently be pumped from the well in the product water. The volume of water containing these exaggerated loadings from the biofouled zone would give an indication of the interstitial volume of the porous and fractured media around the well which has become subjected to biofouling.

In the practice of monitoring a post-quiescent diskharge (i.e., diskharge immediately after a period of passivity in a water well), it is often noted that there are a number of distinctive changes in the floristic composition during the period of diskharge reflective of the particular species being flushed from biofouled zone (i.e., biozone) at any given time.

The sources of the particulates being monitored may be a combination of various layers of biofilm shearing off one after the other from specific biofouled zones close to the well; and imported particulates arriving from the more distant parts of the zone of influence as it gradually extends into the aquifer. Where multiple layer shearing is occurring from a single specific source site, it may be expected that there would be considerable variability in the microbial characteristics of each water sample depending upon the precise composition of the layers shearing within each sample. Where there is a fringe "flushing" effect and the particulates have entered from roughly the same distances from the well screen, the variety and population within various samples should be more consistent. Radical variations in composition between duplicate samples in terms of both population size and composition would indicate a randomized radical shearing of biomasses close to the well (specific source site subject to multiple layered shearing). If there are simultaneous changes in the population size (usually downwards) and the composition of the incumbent flora, there would be a greater probability that the biofouling now extends deeper out into the formation through a number of distinct concentric biozones. Each biozone would be defined by the unique floristic composition and population size recorded.

In either event, the total volume of water pumped prior to the stabilization of the characteristics associated with the producing well would indicate some information on the dimensionality of the biofouled zones. To project this, the primary hypotheses would be that all of the water initially pumped with heightened microbial and/or particulate loadings may be considered to have come from the biofouled zone around the well. In treating the biofouled zone by physical and/or chemical means, the total volume of the interstitial spaces interfacing with the biomasses can be conveniently projected by multiplying the pump rate (pr) by the time of pumping (tp) before the water returned to its more normal background characteristics commonly observed during prolonged periods of pumping of a stabilized well. By this means the biofouled volume (bv) can be projected:

$$bv = pr * tp$$

Once the volume has been calculated, the treatment scenario can be generated. However, where there are a number of distinctive biozones, there may be some need to design different treatments which will function in the most appropriate manner for each of the defined biozones.

To aid in the conceptualization of the size and position of the biofouled zone(s) around the well, a theoretical model (WELLRADI) has been developed. This program projects the extent to which the biofouling would extend around the well (from the midpoint) in the form of either a single cylinder of biomass or as a concentric series of cylinders each encompassing a different type of biomass (e.g., aerobic pseudomonads, anaerobic gas formers, SRBs). This model assumes that all of the biofouling is evenly distributed outside the well screen for the full length of the screen, that there is an even entry of water into the screen, and there is no dilution of water from the biofouled zone with water from beyond the zone of influence. These restrictors distort the value of any calculations to simple relationship functions. For the user, the radius of any zone of biofouling around a well is of interest (e.g., well A has a radius of biofouling of 0.65 meters while well B has a radius of 7.54 meters) and has value in deciding the location of a new well.

Calculation of the radius of biofouling incidence (rbi) may be calculated using the WELLRADI software program in the disk supplementary software. To use the program, the following data needs to be known:

> screen depth, sd
> radius of well screen, ws
> radius of any gravel pack around the screen, gr
> porosity factor for the gravel pack, pg
> porosity factor for the aquifer, pa

In addition, the pr and tp values will need to be known. This data can be presented in metric or U.S. imperial units. The computer program will ask which units are to be used before the calculations will be undertaken. Note the porosity factor is given as the factoral component of medium. If the porosity percentage was 37%, then the porosity factor would be 0.37.

Using WELLRADI software to project a single biofouling cylinder.

Boot up the computer, insert the disk containing the file wellradi.exe. At the prompt, type in wellradi and press ENTER.

B:\>wellradi

There will be a screen prompt asking whether "metric" or "us" units are to be used. Select the appropriate unit style by typing in either of the alternates (note: us standards for U.S. standards) which are desired, press ENTER.

A series of questions is now asked in turn; type in the appropriate answer to each question and press ENTER as each one is typed in. There are seven questions in all:

Enter the screen depth in meters: (12)

Enter the radius of the well in centimeters: (7)

Enter the radius of the gravel pack in meters: (1.25)

Enter the pumping time to clean water, in hours: (3.45)

Enter the pumping rate, in liters per min.: (35)

Enter the porosity of the gravel pack: (0.4)

Enter the porosity of the aquifer: (0.2)

Once the last entry has been made a calculation is undertaken to compute the total volume of water pumped and estimate the average radius of biofouling based upon the interstitial volumes within the gravel pack and affected aquifer. The computer now calculates the biofouled volume of discharged water and the radius of biofouling.

Where the data presented was as shown in the brackets above, the following data were obtained:

Biofouled volume: 7,240 liters

Radius of biofouling: 0.688 meters.

For this well, it would mean that 7,240 liters of water had been generated from the fouling biozones around the well where biofouling was occurring and that these zones extended 0.688 meters away from the vertical mid-point of the well. This information gives the volume of the biofouling which requires treatment and also an impression of the extension of the biofouling into the surrounding formations. **Using WELLRADI Software to Project the Radii of a Multiplicity of Biofouling Cylinders Around a Well.**

During the post-quiescent pumping phase, there may be a number of "signal" events when using the BAQC intensive sampling and analysis scenario. These events occur when there is a radical change in the aggressivity and/or dominance between the various groups of microorganisms within the biofouling. When one group is replaced in dominance by another group in a sampling sequence (e.g., RX of 9, dd of 4 for the IRB-BART™ is replaced by an RX of 5 and a dd of 2), this can reflect the water coming from a different biozone. This event would suggest that the dominance has shifted from an aerobic pseudomonad flora of moderate aggressivity to a more aggressive anaerobic gas-generating flora. Since the latter group occurred in the water pumped from deeper out in the formation, it may also be projected that these latter microorganisms dominated in a biozone outside of the one dominated by the former microbial group. By timing the length of time each group predominated, it becomes possible to project the radii for the various concentric cylindrical biozones. Each radius projected by the WELLRADI software program will be given from the midpoint of the well. The widths of each cylindrical biozone (cbw) may therefore be calculated as the difference between the radius calculated at the initiation of water being pumped of that particular floristic composition (rif) to the conclusion or replacement of the presence of those particular microbial dominance in the water (rxf).

$$cbw = rxf - rif$$

By a careful manipulation of the program, the cbw for the different identified biozones could be calculated.

An example of this is given using the data given below. A well (metric units) had an sd, 5; ws, 10.5; gr, 1.25; pg, 0.37; and a pa, 0.24. An IRB-BART™ reaction RX of 5 was observed in samples taken at between (tp) of 0.1 and 0.75 hours with a pr of 44. Using the WELLRADI program it can be extrapolated that the water influenced by these microorganisms was that which was pumped after the first 264 liters on to the 1,980 liters. A total of 1,716 liters was therefore influenced by these organisms. From the software, it can be calculated that the event occurred between 16.3 and 56.7 cms from the midpoint to form a cylinder width of 40.3 cms. Since the well screen was (ws) 10.5 cms from the midpoint, it may be calculated that this zone ranged beyond the screen from 5.8 to 46.1 cms. Although such a projection is theoretical, it becomes possible to define for the user the proximity of a biofouling event to the well. This also aids in the establishment of a treatment regime to control the biofouling by gaining an appreciation of the relationship of the biofouled zone to the well itself. More importantly in practice, the volume (free liquid interstitial space) which will need to be treated is also calculated. Using these volumes and a knowledge of the biofouling organisms, it becomes more convenient to establish the chemical dosages where this is the elected approach.

A multiplicity of concentric cylinders of biofouling (biozones) may be projected by utilizing the pt values at which the occurrence of a particular dominant group of biofouling organisms initially dominated (tp_i) and concluded the dominance (tp_c). The volume of the affected water can be calculated using the pumped volumes calculated by the WELLRADI at the beginning V_{tpi} and the end V_{tpc} and the volume of the biozone (V_{bz}) determined by the formula:

$$V_{bz} \quad = \quad V_{tpc} \quad - \quad V_{tpi}.$$

In generating treatment systems, it is essential to define the volumes of water associated with the well which are bio-fouled.

Appendix Three

Instructions,
Supplementary Software Package
(supplied separately)

There are three disks in this graphics computer disk video presentation. These disks are 3.5" high density floppy disks to be used on 286 or better MS-DOS compatible computers. It is recommended that these programs will operate adequately with a VGA quality monitor. These presentations (constructed in DrawPerfect v1.1, WordPerfect Corporation, Orem, Utah, fax 801-222-5077) include the various graphic drivers for the different major types of monitor. An option exists for all of the software to be operated through Microsoft Windows if desired. Each of these presentations is self-standing, that means that these programs will operate on a computer that is running in DOS without additional software. It is recommended that a faster response time can be achieved by temporarily importing the material to a hard disk if 3.2 megabytes of memory is available.

DISK CONTENTS

DISK A, includes all of the graphics which have been used in the text but includes additional material which would allow each presentation to become more self-explanatory. All of the material is carried within subdirectory A. To enter the subdirectory, at the DOS prompt, type in cd\A and press ENTER. Each graphics presentation may now be accessed by typing the word SHOW followed by a space and then the key word for the presentation. Upon pressing **ENTER** the program will load. The contents of the various presentations are listed below starting with the key word for that particular set of graphics:

BIOFILM1, these graphics are drawn from Chapter 2. Concepts relate to the early stages in the formation of a biofilm.

BIOFILM2, this section takes the biofilm formation from confluence to plugging (see Chapter 3, The Diagnosis of Biofouling, Microbiology of Plugging in a Water Well).

PLUG1, here the formation of plugging through the generation of a series of concentric biozones around a water well are presented (see Chapter 3, The Diagnosis of Biofouling).

PLUG2, mechanisms by which the biozones can be stressed in order to allow a quantification of the size of biozones is shown (see also Chapter 8, Procedures to Estimate the Degree of Biofouling).

GAL, gives the form, function and life cycle for the stalked bacterium *Gallionella* which is often found associated with iron-related bacterial fouling (see also Chapter 5, Differentiation of Microbial Forms in Biofouling Events, Ribbon Formers).

BACT, presents the various tube-forming iron-related bacteria that are commonly associated with biofouling in water wells and some of the simple field techniques for determining the presence/absence of IRB (see also Chapter 5, Differentiation of Microbial Forms in Biofouling Events, Tube Formers)

DISK B, includes graphics aiding in an understanding as to how the BART™ and MOR™ biodetector tests may be employed to assess the microbial aggressivity in any biofouling event in groundwaters. In addition there are two programs for the file storage and interpretation of the

BART™ data, and the calculation of the diameter of biozones around a well based on the BAQC scenario. All of the material is carried within subdirectory **B**. To enter the subdirectory, at the DOS prompt, type in **cd\B** and press **ENTER**. Each of the graphics presentation may now be accessed by typing the word **SHOW**, followed by a space and then the key word for the presentation. Upon pressing **ENTER**, the program will load. All of these presentations relate to Chapter 9, Monitoring Methodologies for MIF Events. For the two programs relating to file storage of BART™ data and the BAQC, follow the instructions given with the scriptors for those programs below. The contents of the various presentations are listed below starting with the key word for that particular set of graphics:

HOW, gives the basic instructions for charging the biodetectors with water for testing.

HOWTAB, the different technique that has to be used to determine the activity levels of the Total Aerobic Bacteria (TAB) using the TAB biodetector.

TAB, interpretation of the TAB biodetector reaction (DB, Days to Bleaching).

IRB, presentation of the ten reaction patterns which can commonly occur with this test. This test has been designed to detect many of the different forms of iron related bacteria belonging to the CHI group (see Chapter 5, CHI generic groupings). There is an additional reaction (RX 13) which can occur in any of the biodetector series where molds are particularly aggressive.

SLYM, displays the six reactions which can be generated using these biodetectors. Bacteria recovered by this test tend to be those which produce copious amounts of ECPS (i.e., slime formers).

SRB, the sulfate reducing bacteria (see Chapter 5, sulfur bacteria) can be detected by the generation of black sulfides. This display gives the various locations where such reactions can occur. Additionally, there are occasions where there are no SRBs present but a very aggressive anaerobic bacterial growth occurs (reaction X).

FLORPOOL, three reactions are generated in these tests when pseudomonad bacteria are present. Some of these bacteria can cause eye, ear, skin and gastroenteric infections as well as various biofouling and biodegradation (aerobic) activities. These tests have been developed for application in natural and hazardous waters (FLOR) and for recreational waters such as swimming pools and whirlpools (POOL). Since the reaction patterns are similar, both biodetectors are shown together. It should be noted that "thread-like" processes may also be observed (reaction 4) on the POOL biodetector.

COLI, two different coliform test systems are presented which use the (patent pending) gas thimble elevation technique to record the presence of active coliform bacteria in general (G-, total coliforms) or specific (S-, fecal coliform) terms. This technique is suitable for simple presence/absence testing and is described in Chapter 9, coliform testing.

SOFT, describes the various stages in the operation of the BART™-SOFT file storage, sorting and interpretation program. While these graphics give an overview, it is recommended that serious users also read Appendix One.

Note that the following two programs are not graphics programs and may be entered directly from the subdirectory prompt, using the first word printed below for each program and then pressing ENTER.

BART, this program is an .EXE file which allows BART biodetector data to be filed in individual fields for each test. It is essential that the computer clock be set correctly for this program since the days of delay function is calculated using computer generated time. There is built into the program a database which allows the prediction of the possible population once the first *dd* is entered. This program may be transferred to another diskto allow more file space for the individual field. The reader should read Appendix One and view SHOW SOFT before operating the program for the first time.

WELLRADI, a simple program which allows the diameter for the start or conclusion of a particular biofouling event around a water well to be crudely calculated in metric or US units. A series of questions will be asked with respect to the well dimensions, pumping rate, porosity of the pack around the well and the aquifer outside (given as fractions such as 0.41) and the length of time the enhanced fouling was observed issuing from the well. This program will compute the volume of fouled water which issued from the well and the mean projected radius of fouling from the midpoint of the well. Enhanced fouling may cause the discoloration of the water, higher levels of microbial presences (e.g., as measured by reduced dd times in BART™ test) or biofouling-related chemical parameters (such as total iron concentration). It will calculate the diameter and volume of water (from midpoint of the well) to the event being determined. See Chapter 8 and Appendix Two.

Disk C, contains only one show file which includes all of the figures from the test. These figures have been colonized to clarify the major features. These may be viewed by typing SHOW BOOK and pressing enter. It is recommended that this file be stored in a separate subdirectory (such as C) using the same command structure as for the installation of the other disks.

SAVING TO HARD DISK

Where it is expected that these files would be used frequently or it is desirable to have faster accessing, these disks can be transferred to the hard disk where space (2.4 megabytes) is available. It is recommended that each disk be saved in a separate subdirectory (e.g., disk A in subdirectory A; disk B in subdirectory B; disk C in subdirectory C). These subdirectories can be made by going to the hard disk prompt (e.g., C: >) and typing in md\A or md\B respectively and press ENTER. The hard disk will now created an A, B or C subdirectory. Shift the prompt to the drive the disk is in (e.g., A drive) and type **copy *.* C:\A** (or **copy *.* C:\B** or copy *.*C:\C) and press ENTER. The contents of the disk will be copied to the hard disk for future use. To access the hard disk file type **cd\A, cd\B** or **cd\C** as is appropriate and press ENTER. The programs may now be entered. Note that where subdirectories A, B and/or C have already been designated and are in use, a new code system may be used involving no more than eight letters.

Appendix Four

This appendix is composed of items of interest to the reader and is subdivided into two sections:
4.1 Conversion tables
4.2 Sources for microbial culture media

4.1 Conversion Tables

This book has been written predominantly in metric and the following tables are provided for those readers who would wish to use the more traditional units.

Temperature conversion, Centigrade to Fahrenheit:

°C	°F
-17.8	0
-11.1	12
-4.4	24
0	32
5.6	42
7.8	46
11.1	52
15.6	60
20	68
24.4	76
28.9	84
33.3	92
35.6	96
37.8	100
43.0	110
49.0	120
60.2	140

71.4	160
82.6	180

Metric-English Equivalents

Metric unit	Multiplied by	= English Unit
m	3.279	ft
L	0.2642	gal
cm	0.394	in.
kg	2.203	lb
g	0.0353	oz
kPa	0.145	psi
km	0.62137	mile
ha	2.47	acres
km^2	0.386	sq. mile
m^3	1.308	cu. yds

Culture Media Formulations

Most of the microbiological culture media in common use in the analysis of waters and wastewaters is described in the Standard Methods. Additional formulations are available in the various manuals published by the manufacturers. A selection of these manuals in common use is listed below:

DIFCO Manual (Difco Laboratories, Detroit, Michigan 48232 USA)

BBL Manual of Products and Laboratory Procedures (BBL, Division of Becton, Dickinson and Company, Cockeysville, Maryland 21030 USA)

Microbiology Manual (E. Merck, Darmstadt, Germany).

It should be noted that other companies also publish lists of their culture media products and various appropriate methodologies applicable to their product lines and may be an equally appropriate source of information.

Two media which may not appear in the above manuals are listed below (amounts given are per liter):

AA agar medium,

Trypticase peptone	10g
Mannitol	10g
NaCl	1.0g
Nitrofurantoin	0.35g
Crystal violet	0.02g
Agar	14g

WR agar medium,

Ferric ammonium citrate	10g
$MgSO_4$	2.4g
NH_4NO_3	0.5g
K_2HPO_4	0.5g
$NaNO_3$	0.5g
$CaCl_2$	0.15g
Agar	18g

Note: correct pH to 7.4 with 1 N NaOH.

References

Below is a list of references which may provide further information within particular topics. These references are listed as being either:

TB - Textbooks for general reference
RP - Review Papers covering a broad topic
TP - Technical Papers
SP - Scientific Papers
CP - Major Collection of Relevant Papers
PB - Practical Methodologies Book

1. Alford, G., Mansuy, N. and D.R. Cullimore (1989) The Utilization of the Blended Chemical Heat Treatment (BCHT) Process to Restore Production Capacities to Biofouled Water Wells. Proc. Third National Outdoor Action Conference on Aquifer Restoration, Groundwater Monitoring and Geophysical Methods, presented by the Assoc. of Groundwater Scientists and Engineers division of the NWWA. session 1A:229-237 **SP**
2. Allen, M.J. (1980) Microbiology of Ground Water. J. Water Poll. Control Fed. 52(6):1804-1807 **RP**
3. Atlas, R.M. and R. Bartha (1981) Microbial Ecology: Fundementals and Applications. Publ. Addison-Wesley Publishing Co., Reading, Massachusetts, USA, 560 pp. **TB**
4. Balkwill, D.L., Leach,F.R., Wilson, J.T., MacNabb, J.F., and D.C. White (1988) Equivalence of Microbial Biomass Measures Based on Membrane Lipid and Cell Wall Components, Adenosine Triphosphate, and Direct Counts in

Subsurface Aquifer Sediments. Microbial Ecology 16:73-84 **SP**

5. Beloin, R.M., Sinclair, J.L., and Ghiorse, W.C. (1988) Distribution and Activity of Microorganisms in Subsurface Sediments of a Pristine Study Site in Oklahoma. Microbial Ecology 16:85-97 **SP**

6. Bitton, G. and C.P. Gerba (1984) Groundwater Pollution Microbiology. Publ. John Wiley & Sons, New York, USA, 377 pp. **CP**

7. Bitton, G. and S.R. Farrah (1985) Viral Contamination of Groundwater. Rev. Int. Sci. Eau 2(2):31-37 **RP**

8. Bone, T.L. and D.L. Balkwill (1988) Morphological and Cultural Comparison of Microorganisms in Surface Soil and Subsurface Sediments at a Pristine Study Site in Oklahoma. Microbial Ecology 16:49-64 **SP**

9. Bouwer, E.J. and P.L.McCarty (1984) Modelling of Trace Organics Biotransformation in the Subsurface. Ground Water 22(4):433-440 **SP**

10. Boyd, R.F. and B.G. Hoerl (1986) Basic Medical Microbiology, 4th edition. Publ. Little, Brown and Company, Boston, USA. 949 pp. **TB**

11. Brewer, W.S. and J. Lucas (1982) Examination of Probable Causes of Individual Water Supply Contamination. J. Environ. Health 44(6):305-307 **SP**

12. Brock, T.D. and M.T. Madigan (1991) Biology of Microorganisms, 6th edition. Publ. Prentice Hall, Englewood Cliffs, NJ, USA 874 pp. **TB**

13. Buchanon-Mappin, J.M., Wallis, P.M. and A.G. Buchanan (1986) Enumeration and Identification of Heterotrophic Bacteria in Groundwater and a Mountain Stream. Can. J. Microbiol. 32(9):93-98 **SP**

14. Caldwell, B.A. and R.Y. Morita (1988) Sampling Regimes and Bacteriological Test for Coliform Detection in Groundwater. US EPA/600/S2-87/083 3pp **TP**

15. Caldwell, D.E., Kieft, T.L. and D.K. Brannen (1984) Colonization of Sulfide-Oxygen interfaces on Hot Spring Tufa by *Thermothrix thiopara*. J. Geomicrobiol. 3(3):181-189 **SP**

16. Costerton, J.W. and H.M. Lappin-Smith (1989) Behavior of Bacteria in Biofilms. Am. Soc. Microbiol. News 55(12):650-654 **RP**

17. Crane, S.R. and J.A. Moore (1986) Modelling Enteric Bacterial Die-Off: A Review. Water Air Soil Pollut. 27(3-4):411-439 **RP**

18. Cullimore, D.R. and A.E. McCann (1978) The Identification, Cultivation and Control of Iron Bacteria in Ground Water. Publ. in Aquatic Microbiology (ed. Skinner F.A. and J.M. Shewan), Academic Press, New York, USA, 219-261. **RP**

19. Cullimore, D.R. (1981) The Bulyea Experiment (Water Well Pasteurization) Can. J. Water Well Assoc. 7(3):18-21 **TP**

20. Cullimore, D.R. (1983) The Study of a Pseudomonad Infestation in a Well at Shilo, Manitoba. Ground Water 21(5):558-563 **SP**

21. Cullimore, D.R. (1986) editor, IPSCO 1986 Think Tank on Biofilms and Biofouling in Wells and Groundwater Systems. Publ. Regina Water Research Institute, University of Regina, Canada. **RP**

22. Cullimore, D.R. (1987) editor, International Symposium on Biofouled Aquifers: Prevention and Restoration. TPS-87-1. Publ. American Water Resources Association, Bethesda, Maryland, USA, 183 pp. **CP**

23. Daumas, S., Lombart, R. and A. Bianchi (1986) A Bacteriological Study of Geothermal Spring Waters Dating from the Dogger and Trias Period in the Paris Basin. J. Geomicrobiol.4(4):423-434 **SP**

24. Delfino, J.J. and C.J. Miles (1985) Aerobic and Anaerobic Degradation of Organic Contaminants in Florida

Groundwater. Proc. Soil Crop Sci. Soc. Fla. 44:9-14 **SP**
25. Dott, W., Frank, C. and P. Werner (1983) Microbiological Examinations of Groundwater Polluted with Hydrocarbons, 1. Communication: Quantitative and Qualitative Distribution of Bacteria. Zentralbl. Bakteriol. Microbiol. Hyg., Abt. B. 180(1)62-75 **SP**
26. Driscoll, F.G. (1986) Groundwater and Wells, Second Edition, Publ. Johnson Division, St Paul, Minn., USA. 1089pp **TB**
27. Ehrlich, G.G., Godsy, E.M., Goerlitz, D.F. and M.F. Hult (1983) Microbial Ecology of a Creosote-Contaminated Aquifer at St. Louis Park, Minnesota. Dev. Ind. Microbiol. 24:235-246 **SP**
28. Fliermans, C.B. and T.C. Hazen (1990) editors, Proceedings of the First International Symposium on Microbiology of the Deep Subsurface. Publ. Westinghouse Savannah River Company, Aiken, SC, USA, 711 pp **CP**
29. Ford H.W. (1976) Controlling Slimes of Sulfur Bacteria in Drip Irrigation Systems. HortScience 11(2):133-135 **TP**
30. Ford H.W. (1979) The Complex Nature of Ochre. Z.f Kulturtechnik und Flurbereinigung 20:226-232 **SP**
31. Geldreich, E.E. (1986) Control of Microorganisms of Public Health Concern in Water. J. Enviro. Sci. 29:34-37 **RP**
32. Geldreich, E.E. (1989) Drinking Water Microbiology - New Directions toward Water Quality Enhancement. Int. J. Food Micro. 9:295-312 **RP**
33. Geldreich, E.E., Nash, H.D., Spino, D.F. and D.J. Reasoner (1985) Bacterial Colonisation of Point-of-use Water Treatment Devices. J. Am. Water Works Assoc. 77:72-80 **SP**
34. Geldreich, E.E. and E.W. Rice (1987) Occurrence, Significance and Detection of *Klebsiella* in Water Systems. J. Am. Water Works Assoc. 79:74-80 **SP**

35.   Gerba, C.P. (1983) Virus Survival and Transport in Groundwater. Dev. Ind. Microbiol. 24:247-254 **RP**

36.   Ghiorse, W.C. (1984) Biology of Iron- and Manganese- Depositing Bacteria. Ann. Rev. Micro. 38:515-550 **RP**

37.   Grainge, J.W. and E. Lund (1969) Quick Culturing and Control of Iron Bacteria. J. Am. Water Works Assoc. 61(5):242-245 **TP**

38.   Guillemin, F., Henry, P. and L. Monjour (1986) Nitrate Content of Ground Water is Not a Valid Indicator of Fecal Pollution in Rural Sahel Regions. Acta. Trop. 43(2):185-186 **SP**

39.   Hackett,G. and J.H. Lehr (1985) Iron Bacterial Occurrence, Problems, and Control Methods in Water Wells. Publ. National Water Well Association, Dublin, Ohio, USA. **TM**

40.   Haley, M.P. and I.H. Joyce (1984) Modern Instrumental Techniques for Particle Size Evaluation. J. Can. Ceramic Soc. 53:15-20 **SP**

41.   Hallberg, R.O. and R. Martinell (1976) Vyredox - in situ Purification of Ground Water. Ground Water 14(2):88-93 **TP**

42.   Hamilton, W.A. (1985) Sulphate-Reducing Bacteria and Anaerobic Corrosion. Ann. Rev. Microbiol. 39: 195-217 **RP**

43.   Harley, J.P. and L.M. Prescott (1990) Laboratory Exercises in Microbiology.   Publ. Wm. C. Brown publishers, Dubuque, IA, USA. 416 pp. **PB**

44.   Harvey, R.W., Smith, R.L. and L. George (1984) Effect of Organic Contamination upon Microbial Distributions and Heterotrophic Uptake in a Cape Cod, Mass. Aquifer. Appl. Environ. Microbiol. 48(6):1197-1202 **SP**

45.   Hasselbarth, U. and D. Ludermann (1972) Biological Incrustation of Wells due to Bacteria. Water Treat. Exam.

21:20-29 **SP**

46. Hattori, T. (1988) The Viable Count: Quantitative and Environmental Aspects. Publ. Science Tech Publishers, Madison, Wisc., USA & Springer-Verlag, New York, USA. 88pp **RP**

47. Hirsch, P. and E. Rades-Rohkohl (1983) Microbial Diversity in a Groundwater Aquifer in Northern Germany. Dev. Ind. Microbiol. 24:183-200 **RP**

48. Howsam, P. (1988) Biofouling in Wells and Aquifers. J. Inst. Water and Environmental Management 2(2):209-215 **SP**

49. Howsam,P. and S.F. Tyrrel (1989) Diagnosis and Monitoring of Biofouling in Enclosed Flow Systems - Experience in Ground-Water Systems. Jour. Biofouling 1:343-351 **SP**

50. Howsam, P. (1990) editor, Microbiology in Civil Engineering, FEMS symposium No. 59. Publ. E. & F.N. Spon., London, UK. 382 pp **CP**

51. Howsam, P. (1990) editor, Water Wells Montoring, Maintenance, Rehabilitation. Publ. E. & F.N. Spon., London, UK. 422 pp **CP**

52. Jackson, R.E. (1982) The Contamination and Protection of Aquifers. Nat. Resour., 18(3):2-6 **RP**

53. Jacobs, N.J., Zieglers, W.L., Reed, F.C., Stukel, T.A., and E.W. Rice (1986) Comparison of Membrane Filter, Multiple-fermentation-tube, and presence-absence techniques for detecting total coliforms in small community water systems. Appl. Environ. Microbiol. 51:1007-1012 **SP**

54. Jarvis, W.R., White, J.W., Munn, Van P., Mosser, J.L., Emori, T.G., Culver, D.H.. Thornsberry, C. and J. Hughes (1985) Nosocomial Infection Surveillance, 1983. US, CDC Surveillance Summaries 33(#2SS):9SS-21SS **RP**

55. Keswick, B.H., Wang, D.S. and C.P. Gerba (1982) The Use of Microorganisms as Groundwater Tracers: A

review. Ground Water 20(2):142-149 **RP**
56. Lindblad-Passe, A. (1988) Clogging Problems in Groundwater Heat Pump Systems in Sweden. Water Sc. Technol. 20(3):133-140 **SP**
57. Lueschow, L.A. and K.M. Mackenthum (1962) Detection and Enumeration of Iron Bacteria in Municipal Water Supplies. Jour. AWWA 54:751-756 **SP**
58. Lynch, J.M. and J.E. Hobbie (1988) Microorganisms in Action: Concepts and Applications in Microbial Ecology. Publ. Blackwell Scientific Publications, Palo Alto, Calif, USA. 363 pp. **TB**
59. Marxsen, J. (1988) Investigations into the Number of Respiring Bacteria in Groundwater from Sandy and Gravelly Deposits. Microbial Ecolgy 16:65-72 **SP**
60. Molozhavaya, E.T., Chugunikhina, N.V. and M.I. Afanasieva (1979) GIG. Sanit. 8:22-26 **SP**
61. Ogunseitan, O.A., Tedford, E.T., Pacia, D., Sirotkin, K.M. and G.S. Salver (1987) Distribution of Plasmids in Groundwater Bacteria. J. Ind. Microbiol. 1(8):311-317 **SP**
62. Prescott, L.M., Harley, J.P. and D.A. Klein (1990) Microbiology. Publ. Wm. C. Brown publishers, Dubuque, IA, USA. 883 pp. **TB**
63. Pye, V.I. and R. Patrick (1983) Ground Water Contamination in the United States. Science 221(#4612):713-718 **RP**
64. Reasoner, D.J. (1983) Microbiology of Potable Water and Groundwater. J. Water Pollut. Control Fed. 55(6):891-895 **RP**
65. Roa, C.S.G. (1970) Occurence of Iron Bacteria in the Tube Well Water Supply of Howrah. Environ. Health 12:273-280 **SP**
66. Rodina, A.G. (1972) Methods in Aquatic Microbiology (translated by R.R. Colewell and M.S. Zambruski). Publ. University Park Press, Baltimore,

Maryland, USA,461 pp. **PB**
67. Schweisfurth, R. (1986) Biofiltration in Drinking Water Treatment Publ in Biotechnology 8 chapter 11 (editors, Rehm, H.J. and G. Reed) 399:423
68. Sinton, L.W. (1984) The Macroinvertebrates in a Sewage-Polluted Aquifer. Hydrobiologia. 119(3):161-169 **RP**
69. Smith, G.A., Nickels, J.S., Kerger, B.D., Davis J.D., Collins, S.P., Wilson, J.T., MacNabb, J.F. and D.C. White (1986) Quantitative Characterization of Microbial Biomass and Community Structure in Subsurface Material: a Procaryotic Consortial Responsive to Organic Contamination. Can. J. Microbiol. 32:104-111 **SP**
70. Smith, S.A. (1982) Culture Methods for the Enumeration of Iron Bacteria from Water Well Samples - a critical literature review. Ground Water 20(4)482-485 **TP**
71. Smith, S.A. (1985) Analyzing Groundwater Quality for POU Treatment. Water Tech. May, 27-34
72. Smith, S.A. (1992) Monitoring Methods for Iron/Manganese Biofouling Publ. America Water Works Association Research Foundation and the AWWA, Denver, CO, USA, 149 pp (in press) **RP**
73. Smith, S.A. and O.H. Tuovinen (1985) Environmental Analysis of Iron-Precipitating Bacteria in Ground Water and Wells. Groundwater Monitoring Rev. 5(4):45-52 **SP**
74. Smolenski, W.J. and J.M. Sulfita (1987) Biodegradation of Cresol Isomers in Anoxic Aquifers. Appl. Environ. Microbiol. 53(4):710-716 **SP**
75. Stetzenbach, L.D., Kelley, L.M. and N.A. Sinclair (1986) Isolation, Identification, and Growth of Well-Water Bacteria. Ground Water 24(1):6-10 **SP**
76. Thorn, P.M., and R.M. Ventullo (1988) Measurement of Bacterial Growth Rates in Subsurface Sediments Uisng the Incorporation of Tritiated Thymidine into DNA. Microbial Ecology 16:3-16 **SP**

77. Wallis, P.M. (1983) Organic Biogeochemistry at a Mountain Coal Mine. Geomicrobiology 3(1):49-78 **SP**
78. Walsh, E. and R. Mitchell (1972) A pH-dependent Succession of Iron Bacteria. Environ. Sci. Technol. 6:809-812 **SP**
79. Webster, J.J., Hampton, G.J., Wilson, J.T., Ghiorse, W.C. and F.R. Leach (1985) Determination of Microbial Cell Numbers in Subsurface Samples. Ground Water 23(1):17-25.
80. Wellings, F.M. (1982) Viruses in Groundwaters. Environ. Int. 7(1) ISBN 0-08-029392-1 **RP**
81. West, Candida Cook (1990) Transport of Macromolecules and Humate Colloids Through a Sand and a Clay Amended Sand Laboratory Column. US EPA/600/S2-90/020 8pp **SP**
82. Wilson, J.T., MacNabb, J.F., Balkwill, D.L. and W.C. Ghiorse (1983) Enumeration and Characterization of Bacteria Indigenous to a Shallow Water-Table Aquifer. Ground Water 21(2):134-142 **SP**
83. Wolfe, R.S. (1958) Cultivation, Morphology and Classification of the Iron Bacteria. J. Am. Water Works Assoc. 50(9):1241-1249 **SP**
84. Yates, M.V. (1990) The Use of Models for Granting Variances from Mandatory Disinfection of Ground Water Used as a Public Resource. US EPA/600/2-90/010 13pp **TP**

Selected Additional Readings

The following is a list of texts which may be useful for routine use in the general area of groundwater microbiology:

Standard Methods for the Examination of Water and Wastewater, 17th edition. Edited by L.S.Clesceri, et al. American Public Health Association, 1989. ISBN 0-87553-161-X 1624 pp.

Standard Methods for the Examination of Water and Wastewater, Supplement to the 17th edition. Edited by A.E.Greenberg, et al. American Public Health Association, 1991. ISBN 0-87553-201-2 150 pp.

Nalco Water Handbook, Second edition. Edited by F. Kemmer. McGraw-Hill 1988 ISBN 0-07-04587-3 1019 pp.

1992 ASTM Standards: Section 11, Water and Environmental Technology volume 11.02 Water II ASTM Standards D-15506-03 1992 ISBN 0-8031-1729-9 714 pp.

Drinking Water Microbiology. G.A.McFeters Springer-Verlag, 1990 ISBN 0-387-97162-9 502 pp.

The Merck Manual of Diagnosis and Therapy, 15th edition Edited by R. Berkow and A. Fletcher, Merck 1987 ISBN 09-11910-06-09 2696 pp.

Standard Handbook of Hazardous Waste Treatment and Disposal by H.M. Freeman. McGraw-Hill 1989 ISBN 0-07-022042-5 926 pp.

Wastewater Engineering: Treatment, Disposal, Reuse Third Edition by Metcalf & Eddy, Inc. McGraw-Hill 1991 ISBN 0-07-041690-7 1334 pp.

Laboratory Exercises in Microbiology J.P.Harley and L.M.Prescott W.C. Brown 1990 ISBN 0-697-03005-9 416 pp. (*)

Microbiology L.M.Prescott, J.P.Harley and D.A.Klein W.C. Brown 1990 ISBN 0-697-00246-2 882 pp text.

It should be noted that any person performing microbiology examinations should be confident in the art. Persons undertaking microbiological studies should minimally have an understanding equivalent to the practical aspects of the text designated by an asterisk (*) above.

Index

DATE DUE

GAYLORD